Communications
in Computer and Information Science   845

Commenced Publication in 2007
Founding and Former Series Editors:
Alfredo Cuzzocrea, Xiaoyong Du, Orhun Kara, Ting Liu, Dominik Ślęzak,
and Xiaokang Yang

More information about this series at http://www.springer.com/series/7899

Yee Ling Boo · David Stirling
Lianhua Chi · Lin Liu
Kok-Leong Ong · Graham Williams (Eds.)

Data Mining

15th Australasian Conference, AusDM 2017
Melbourne, VIC, Australia, August 19–20, 2017
Revised Selected Papers

 Springer

Editors
Yee Ling Boo
RMIT University
Melbourne, VIC
Australia

David Stirling
University of Wollongong
Wollongong, NSW
Australia

Lianhua Chi
La Trobe University
Melbourne, VIC
Australia

Lin Liu
School of Information Technology
  and Mathematical Sciences
University of South Australia
Adelaide, SA
Australia

Kok-Leong Ong
La Trobe University
Melbourne, VIC
Australia

Graham Williams
Microsoft Pty Ltd
Singapore
Singapore

ISSN 1865-0929   ISSN 1865-0937 (electronic)
Communications in Computer and Information Science
ISBN 978-981-13-0291-6   ISBN 978-981-13-0292-3 (eBook)
https://doi.org/10.1007/978-981-13-0292-3

Library of Congress Control Number: 2018940154

Printed on acid-free paper

This Springer imprint is published by the registered company Springer Nature Singapore Pte Ltd.
part of Springer Nature
The registered company address is: 152 Beach Road, #21-01/04 Gateway East, Singapore 189721, Singapore

Preface

We are pleased to present the proceeding of the 15th Australasian Data Mining Conference (AusDM 2017). The Australasian Data Mining (AusDM) Conference series began in 2002 as a Workshop and has grown each year since. The series was initiated by Dr. Simeon Simoff (then Associate Professor, University of Technology, Sydney), Dr. Graham Williams (then Principal Data Miner, Australian Taxation Office, and Adjunct Professor, University of Canberra), and Dr. Markus Hegland (Australian National University). It continues today with Professors Simoff and Williams chairing the Steering Committee.

The Australasian Data Mining Conference is devoted to the art and science of intelligent data mining: the meaningful analysis of (usually large) data sets to discover relationships and present the data in novel ways that are compact, comprehendible, and useful for researchers and practitioners. For the 15th year, the conference continued bringing together data mining researchers and data science practitioners to share and learn of research and progress in the local context and to hear of new breakthroughs in data mining algorithms and their applications.

The conference has also developed into a premier venue for postgraduate students in data mining and data science to come together each year to present their research, to network with other students and researchers, and to showcase their ideas with industry. In the past few years, AusDM has gone from a single-track conference to a dual-track conference covering both research and application aspects of data science. The conference organizers are also keen to experiment each year, partnering with various conferences through co-location, or on its own at regional areas of Australia. AusDM 2017 continued this spirit and this year we joined the premier international AI conference (IJCAI) as one of its workshops out of the 30+ workshops being co-located with IJCAI.

Despite co-locating with such a large-scale conference, AusDM continues to attract a good number of submissions from its two tracks. A total of 32 papers were received. One of the papers was disqualified for not meeting submission requirements and the remaining 31 papers were reviewed by at least three reviewers. The review process was double-blinded and where review outcomes were unclear from the initial three reviewers, additional reviewers were enlisted in order to reach a clear accept/reject outcome. In addition, all review comments that were returned were checked by the respective program chairs to ensure they were constructive and valid responses were given so that papers could be improved before publication or in the case of the rejected papers, could be improved for future submission. Of the 31 peer-reviewed papers, a total of 17 papers were accepted with 11 from research track (out of a total of 22) and six application track papers (out of a total of nine). The overall acceptance rate for this year's conference sits at 55%, which is an exception and AusDM will revert back to its long-term acceptance rate of below 40% going forward when it is located on its own in 2018.

In addition to the paper presentations, the chairs would like to thank SAS Australia for sponsoring the keynote speaker, and the various other sponsors for putting together the conference memento for AusDM 2017. The Organizing Committee would also like to thank Springer CCIS and the Editorial Board for their acceptance to publish AusDM 2017 as a post-acceptance proceedings, which we believe will greatly improve the exposure of the research reported in this conference and we look forward to working closely with Springer CCIS. Lastly, we like to thank all conference delegates for being part of the program this year and we hope you enjoyed the conference program of 2017.

January 2018

Yee Ling Boo
David Stirling
Lianhua Chi
Lin Liu
Kok-Leong Ong
Graham Williams

Organization

Conference Chairs

Kok-Leong Ong   La Trobe University, Australia
Graham Williams   Microsoft, Singapore

Program Chairs (Research Track)

Lin Liu   University of South Australia, Australia
Yee Ling Boo   RMIT University, Australia

Program Chairs (Application Track)

Lianhua Chi   IBM Research, Australia
David Stirling   University of Wollongong, Australia

Tutorial Chair

Andrew Stranieri   Federation University, Australia

Steering Committee Chairs

Simeon Simoff   University of Western Sydney, Singapore
Graham Williams   Microsoft, Singapore

Steering Committee

Peter Christen   Australian National University, Australia
Ling Chen   University of Technology, Australia
Zahid Islam   Charles Sturt University, Australia
Paul Kennedy   University of Technology, Australia
Jiuyong (John) Li   University of South Australia, Australia
Kok-Leong Ong   La Trobe University, Australia
Yanchang Zhao   CISRO Data61, Australia
Andrew Stranieri   Federation University, Australia
Richi Nayak   Queensland University of Technology, Australia
Dharmendra Sharma   Canberra University, Australia

Honorary Advisors

John Roddick   Flinders University, Australia
Geoff Webb   Monash University, Australia

Program Committee

Research Track

Adil Bagirov	Federation University, Australia
Jie Chen	University of South Australia, Australia
Xuan-Hong Dang	University of California at Santa Barbara, USA
Ashad Kabir	Charles Sturt University, Australia
Wei Kang	University of South Australia, Australia
Yun Sing Koh	University of Auckland, New Zealand
Cheng Li	Deakin University, Australia
Gang Li	Deakin University, Australia
Kewen Liao	Swinburne University of Technology, Australia
Brad Malin	Vanderbilt University, USA
Qinxue Meng	University of Technology Sydney, Australia
Veelasha Moonsamy	Utrecht University, The Netherlands
Muhammad Marwan Muhammad Fuad	Aarhus University, Denmark
Quang Vinh Nguyen	Western Sydney University, Australia
Jianzhong Qi	University of Melbourne, Australia
Azizur Rahman	Charles Sturt University, Australia
Md Anisur Rahman	Charles Sturt University, Australia
Jia Rong	Victoria University, Australia
Grace Rumantir	Monash University, Australia
Flora Dilys Salim	RMIT University, Australia
Dharmendra Sharma	University of Canberra, Australia
Glenn Stone	Western Sydney University, Australia
Xiaohui Tao	University of Southern Queensland, Australia
Dhananjay Thiruvady	Monash University, Australia
Truyen Tran	Deakin University, Australia
Dinusha Vatsalan	Australian National University, Australia
Sitalakshmi Venkatraman	Melbourne Polytechnic, Australia
Lei Wang	University of Wollongong, Australia
Guandong Xu	University of Technology Sydney, Australia
Ji Zhang	University of Southern Queensland, Australia
Rui Zhang	University of Melbourne, Australia

Application Track

Alex Antic	PricewaterhouseCoopers, Australia
Chris Barnes	University of Canberra, Australia
Rohan Baxter	Australian Tax Office, Australia
Nathan Brewer	Department of Human Services, Australia
Neil Brittliff	University of Canberra, Australia
Adriel Cheng	Defence Science and Technology Group, Australia
Tania Churchill	AUSTRAC, Australia

Hoa Dam	University of Wollongong, Australia
Klaus Felsche	C21 Directions, Australia
Markus Hagenbuchner	University of Wollongong, Australia
Edward Kang	Australian Passport Office, Australia
Luke Lake	DHS, Australia
Jin Li	Geoscience Australia, Australia
Balapuwaduge Sumudu Udaya Mendis	Australian National University, Australia
Tom Osborn	University of Technology Sydney, Australia
Martin Rennhackkamp	PBT Group, Australia
Goce Ristanoski	CSIRO Data61, Australia
Nandita Sharma	Australian Taxation Office, Australia
Chao Sun	University of Sydney, Australia
Junfu Yin	University of Technology Sydney, Australia
Ting Yu	Commonwealth Bank of Australia, Australia
Yanchang Zhao	CISRO Data61, Australia

Sponsors

Keynote Sponsor

SAS Australia Pty Ltd

Other Sponsors

La Trobe University
RMIT University
University of South Australia
University of Wollongong
IBM Australia

Contents

Clustering and Classification

Similarity Majority Under-Sampling Technique for Easing Imbalanced Classification Problem

Jinyan Li[1($\boxtimes$)], Simon Fong[1($\boxtimes$)], Shimin Hu[1], Raymond K. Wong[2],
and Sabah Mohammed[3]

[1] Department of Computer and Information Science, University of Macau,
Taipa, Macau SAR, China
{yb47432, ccfong, yb72021}@umac.mo
[2] School of Computer Science and Engineering,
University of New South Wales, Sydney, NSW, Australia
wong@cse.unsw.edu.au
[3] Department of Computer Science, Lakehead University, Thunder Bay, Canada
mohammed@lakeheadu.ca

Abstract. Imbalanced classification problem is an enthusiastic topic in the fields of data mining, machine learning and pattern recognition. The imbalanced distributions of different class samples result in the classifier being over-fitted by learning too many majority class samples and under-fitted in recognizing minority class samples. Prior methods attempt to ease imbalanced problem through sampling techniques, in order to re-assign and rebalance the distributions of imbalanced dataset. In this paper, we proposed a novel notion to under-sample the majority class size for adjusting the original imbalanced class distributions. This method is called Similarity Majority Under-sampling Technique (SMUTE). By calculating the similarity of each majority class sample and observing its surrounding minority class samples, SMUTE effectively separates the majority and minority class samples to increase the recognition power for each class. The experimental results show that SMUTE could outperform the current under-sampling methods when the same under-sampling rate is used.

Keywords: Imbalanced classification · Under-sampling · Similarity measure
SMUTE

1 Introduction

Classification is a popular data mining task. A trained classifier is a classification model which is inferred from training data that predicts the category of unknown samples. However, most of current classifiers assume that the distribution of dataset is balanced. Practically, most datasets found in real life are imbalanced. This gives rise to weakening the recognition power of the classifier with respect to minority class, and probably overfitting the model with too much training samples from majority class.

In essence, the imbalanced problem which degrades the classification accuracy is rooted at the imbalanced dataset, where majority class samples outnumbers those of the

© Springer Nature Singapore Pte Ltd. 2018
Y. L. Boo et al. (Eds.): AusDM 2017, CCIS 845, pp. 3–23, 2018.
https://doi.org/10.1007/978-981-13-0292-3_1

minority class in quantity. E.g. the ratios of majority class samples and minority class samples at 20:1, 100:1, 1000:1, and even 10000:1 [1] are not uncommon. The reason for attracting the researcher's attention is that, in most cases the minority class is the prediction target which is of interest while the massive majority class samples are ordinary. The imbalanced classification problems often appear naturally in real-life applications, such as in bioinformatics dataset analysis [2], forecasting nature disasters [3], image processing [4] as well as assisting diagnosis and treatment through biomedical and health care datasets [5].

Since conventional classifiers are designed to learn the relation between input variables and target classes, without regards to whichever class the samples come from. Feeding imbalanced binary class dataset to the model building algorithm, the majority samples will bias the classifier with a tendency of overfitting to the majority class samples and neglecting minority class samples. At the end, since the majority class samples dominate a large proportion in the training dataset, the classification model will still appear to be very accurate when being validated with the same dataset which contains mostly the majority samples for which the model was trained very well. However, when the classifier is being tested with the emergence of rare instances, the accuracy drops sharply. When such model is being used in critical situations, such as rare disease prediction, disaster forecast or nuclear facility diagnosis, the insensitivity of the trained model for accurately predicting the rare exceptions would lead to grave consequence.

The drawback of the classification model is due to the lack of training with the few amount of rare samples available. When the model is tested with fresh samples from the minority class, it becomes inadequate. Knowing that "accuracy" is unreliable in situation like this, prior researchers proactively adopted other evaluation metrics to replace or supplement accuracy in order to justly assess the classification model and the corresponding rebalancing techniques. These metrics include AUC/ROC [6], G-mean [7], Kappa statistics [8], Matthews correlation coefficient (MCC) [9], and F-measure [10], etc. In general, researchers tried to solve the imbalanced problem of classification by re-distributing the data from the other major and minor classes through sampling techniques in the hope of improving the classifiers. One common approach is to over-sample more instances from the minor class, even artificially.

In this paper, we propose an alternative and novel under-sampling method, namely Similarity Majority Under-sampling Technique (SMUTE) to ease imbalanced problem as a pre-processing mechanism for reducing the imbalanced ratio in the training dataset. It adopts a filter strategy to select the majority class samples which are shown to work well in combination with the existing minority class samples. It works by referring to the similarity between the majority class samples and minority class samples, then it screens off the majority class samples which are very similar to those minority class simples, according to the given under-sampling rate, in order to reduce the imbalanced ratio between two classes. Firstly, it calculates the similarity of each majority class samples and its surrounding minority class samples. Then each majority class will obtain a value, which is the sum of a given number of the most similar minority class samples' similarity to each majority class sample. Sort these majority class samples by their sum similarity from small to large. Finally, the algorithm retains a give number of majority class samples (e.g. top k) through a filtering approach. This method could effectively segregate majority class samples and minority class samples in data space and maintain high distinguishing degree between each class, in an effort to keep up with the discriminative power and high classification accuracy.

The remaining paper is organized as follows. Some previous approaches and papers for solving imbalanced problem are reviewed in Sect. 2. In Sect. 3, we elaborate our proposed method and the process. Then, the data benchmark, comparison algorithms, our experiment and results are demonstrated in Sect. 4. Section 5 summarizes this paper.

2   Related Work

As above introduced that imbalanced classification is crucial problem to which effective solutions are in demand. Since the conventional classification algorithms were not originally designed to embrace training from imbalanced dataset, it triggered a series of problems, due to overfitting the majority class data and underfitting the minority data. These problems include Data Scarcity [11], Data Noise [12], Inappropriate Decision Making and Inappropriate Assessment Criteria [13].

For overcoming this imbalanced data problem, current methods can be broadly divided into data level and algorithm level. Previous researchers proposed that there are four main factors for tackling imbalanced classification problem. They are (i) training set size, (ii) class priors, (iii) cost of errors in different classes, and (iv) placement of decision boundaries [14, 15]. The data level methods adopt resampling techniques to re-adjust the distribution of imbalanced dataset. At the algorithm level, the conventional classification algorithms are modified to favour the minority class samples through assigning weights on samples that come from different classes or ensemble techniques where the candidate models that are trained with minority class data are selected more often.

Prior arts suggested that rebalancing the dataset at the data level, by pre-processing is relatively simpler and as effective as biasing imbalanced classification [16]. Hence, sampling methods have been commonly used for addressing imbalanced classification by redistributing the imbalanced dataset space. Under-sampling reduces the number of majority class samples and Over-sampling increases the amount of minority class samples. These two sampling approaches are able to get even the imbalanced ratio. However, there is no golden rule on how much exactly to over-sample or under-sample so to achieve the best fit. An easy way is to simply and randomly select majority class samples for downsizing and likewise for repeatedly upsizing minority class samples, randomly. Random under-sampling will lose important samples by chances, and inflating rare samples without limit will easily cause over-fitting too. Synthetic Minority Over-sampling Technique (SMOTE) [17] is one of the most popular and efficient over-sampling methods in the literature. Each minority class sample mimics about several of its neighbour minority class samples to synthesise new minority class samples, for the purpose of rebalancing the imbalanced dataset. The biggest weakness of this method is that the synthesized minority class samples may coincide with the surrounding sample of majority class sample [15]. For this particular weakness, researchers invented a number of modifications, extending SMOTE to better versions: for example, Borderline SMOTE [18], MSMOTE [19] and etc.

Fundamentally, over-sampling will dilute the population of the original minority class samples by generating extra synthetic samples. On the other hand, eliminating some majority class samples by under-sampling helps relieve the imbalanced classification problems too. It is known that random under-sampling could result in dropping some meaningful and representative samples though they are in the majority class. Instead of doing it randomly, Kubat and Matwin adopted one-side under-sampling to remove the noise samples, boundary samples and redundant samples in majority class to subside the imbalanced ratio [20]. The other researchers obtain the balanced number of support vectors by pruning the support vectors of majority class to increase the identification of minority class samples [21]. Some researchers also adopted one-class samples as training dataset to replace the whole dataset and avoid the imbalanced problem [22]. Estabrooks and Japkowic concurrently used over-sampling and under-sampling with different sampling rates to obtain many sub-classifiers, like an ensemble method. The sub-classifiers are then integrated by the frame of mixture-of-experts in the following step [23]. The experimental results showed that this method is much better than the other ensemble methods. Balance Cascade [13] is a classical under-sampling method. Through iteration strategy, it is guided to remove the useless majority class samples gradually.

To sum up, despite the fact that sampling techniques can potentially solve the imbalanced distribution of imbalanced dataset, over-sampling techniques may dilute the minority class samples and under-sampling techniques may remove some important information of majority class samples [24]. Therefore, these methods have limited effect for handling imbalanced classification.

Ensemble learning and Cost-sensitive learning are two core techniques at algorithm level for solving imbalanced problem. They work by assigning different weights or votes or further iterations to bias the ratio, while conventional methods concern about increasing size of the minority class samples.

Ensemble learning gathers a number of base classifiers and then it adopts some ensemble techniques to incorporate them to enhance the performance of classification model. Boosting and Bagging [25], are the most frequently used approaches. AdaBoosting is a typical construct in boosting series methods. It adaptively assigns different and dynamic weights to each sample in iterations to change the tendentiousness of classifier [26]. Bagging implements several variations of sub-classifiers to promote the performance. These sub-classifiers classify repeatedly using the re-sampled dataset. A winning classifier which is most voted would be selected to produce the final results after several rounds of voting.

A lot of research works are focused on over-sampling the minority class samples be it at the data level or tuning up the bias at the algorithm level. It was supposed that properly recognizing the minority class samples is more valuable than the majority class samples. The belief is founded on the consequence that misclassifying any minority class sample would often need to pay a high price in critical applications. Cost-sensitive learning followed this basic idea to assign different costs of misclassified class. Besides attempts to pushing down the misclassification rate, during the training of a cost-sensitive classifier, the classifier will be forced to boost a higher recognition rate for minority class samples, since keeping the total cost at the lowest is one of the optimization goals of this learning model.

In our new design, called SMUTE which will be described as follow, the integrity of the minority class samples as well as their population size would be left untouched. This is principled on the condition that the minority class samples are better to be preserved as original as they are, in such a way that no more or no less amount should be intervened in them. Hence an authentic classifier which is trained right from the original minority class samples, would offer the purist recognition power. Analogous to Occam's razor theory, the best predictive model might be the one that is trained with just the right amount of training samples. Different from most of the popular class rebalancing techniques reviewed above, SMUTE manipulates only at the majority class samples, repopulating those majority instances which are found to be compatible (or similar) to the minority instances.

3   Similarity Majority Under-Sampling Technique

In the data space of an imbalanced dataset, majority class samples occupy most of it. The inequity causes the classifier to be insufficiently trained for minority class samples and the overwhelming majority class samples interfere the identifying power of the minority class samples. Consequently, the classifiers will bias majority class samples and it suffers a pseudo high accuracy if it were tested with imbalanced dataset again. Over-sampling techniques reverse the bias of classifiers through synthesizing new minority class samples. However, the data space will become more crowded, there might even be some overlaps between these samples that give rise to confusion to the training. Essentially, the single over-sampling techniques increase the computational cost of training because extra samples are synthesized, added, increasing the overall training data size, but the discrimination between samples are blurred.

Under-sampling is another reasonable approach to reduce the disproportion between the two classes. Some under-sampling methods are introduced and used hereinbefore, such as instance selection and clustering method. The art of under-sampling is how to reduce the majority class samples, in a way that the effective distinction between samples from different classes is sharpened, while ensuring the data space does not get congested but the class samples are well distributed closely according to the underlying non-linear relations resembled in the classifier. In SMUTE, majority class samples are selected based on how "compatible" the majority samples to the minority samples are, keeping the minority samples intact. Similarity measure is used here as a compatibility check which calculates the similarity degree between two data point in multi-dimensional space. Calculation methods vary for similarity measure. The most common measure is correlation which has been widely used in the similarity measure adheres to four principles: (1) the similarity of their own is 0; (2) the similarity is a non-negative real number quantified as the distance apart; (3) Symmetry, if the similarity from A to B is equal to the similarity from B to A. (4) Triangular rule: the sum of both sides is greater than the third side of the similarity triangle.

Two steps of computation are in the proposed Similarity Majority Under-sampling Technique.

1. Each majority class sample calculates the distances pairing between itself and each of K minority class samples, and sum up these distances to a similarity score.
2. Given an under-sampling rate, [0, 1], select a subset of majority class samples which have the top percentage of high similarity scores; the disqualified samples (which have relatively low similarity scores) are discarded.

In our experiment, eight common methods of calculating similarity are tried, together with 20 equidistant different under-sampling rates ranging from 100% to 0.5% to rebalance the imbalanced datasets.

For a thorough performance assessment, 100 binary class imbalanced datasets from KEEL [27] are used for testing SMUTE with different versions of similarity. A standard C4.5 Decision Tree is used as the base classifier, which is subject to 10-cross validation method for recording the classification performances. For each dataset, each similarity measure method, and at each under-sampling rate will be repeatedly run 10 times before averaging them to a mean value. In addition to accuracy, Kappa statistic [28, 29] is chosen as the main evaluation metric because it indicates how reliable the accuracy the classifier in terms of generalizing its predictive power on other datasets.

The eight similarity measures are Euclidean Distance [30], Manhattan Distance [30], Chebyshev Distance [30], Angle Cosine [31], Minkowski Distance [32], Correlation coefficient [33], Hamming Distance [34] and Jaccard similarity coefficient [35].

Euclidean Distance is the distance between two points in Euclidean space. Equation (1) depicts the distance between two n-dimensional vectors in Euclidean space. The smaller distance between the two samples means they have greater similarity.

$$distance = \sqrt{\sum\nolimits_{i=1}^{n} (x_i - y_i)^2} \tag{1}$$

Manhattan Distance is also called city block distance. In real life, the distance between the two points may not be line-of-sight distance, for there are buildings and obstacles in between them. Equation (2) represents the distance of two n-dimensional vectors in Manhattan Distance. The shorter the distance between the two samples means they are more similar.

$$distance = \sum\nolimits_{i=1}^{n} |x_i - y_i| \tag{2}$$

Chebyshev Distance is also named L∞ measure. The distance of two n-dimensional vectors is the maximum value of the absolute value of each coordinate value. Shorter the distance means higher the similarity. Equation (3) is the formula for Chebyshev Distance.

$$distance = \lim_{k \to \infty} \left(\sum\nolimits_{i=1}^{n} |x_i - y_i|^k \right)^{1/k} \tag{3}$$

Angle Cosine. The Cosine similarity is independent of the amplitude of the vector, only in relation to the direction of the vector. The Angle Cosine is in the range [−1, 1]. The larger the Angle Cosine value means the narrower the angle and the greater similarity between the two vectors, vice versa. When the direction of the two vectors coincides, the Angle Cosine takes the maximum value 1, and when the direction of the two vectors is exactly at opposite, the Angle Cosine takes the minimum value −1.

$$\cos(\theta) = \frac{\sum_{i=1}^{n} x_i y_i}{\sum_{i=1}^{n} x_i^2 \sum_{i=1}^{n} y_i^2} \tag{4}$$

Minkowski Distance can be interpreted as various distance definitions. Equation (5) is the Minkowski distance of two n-dimensional vectors. When $p = 1$, it is Manhattan Distance; $p = 2$ is Euclidean Distance; and when $p \to \infty$, it becomes the formula of Chebyshev distance.

$$distance = \sqrt[p]{\sum_{i=1}^{n} |x_i - y_i|^p} \tag{5}$$

Correlation coefficient is defined as Eq. (6). It is a measure of the correlation between the random variables X and Y, and the correlation coefficient is capped at −1 and 1. The greater absolute value of the correlations indicates the higher correlation between X and Y. When X is linearly related to Y, the correlation coefficient is 1 or −1, which respectively means positive linear correlation and negative linear correlation.

$$p_{XY} = \frac{n \sum xy - \sum x \sum y}{\sqrt{n \sum x^2 - (\sum x)^2} - \sqrt{n \sum y^2 - (\sum y)^2}} \tag{6}$$

Hamming Distance. The Hamming distance between the two equal length strings $s1$ and $s2$ is defined as the minimum number of times that one of them becomes the other. For example, the hamming distance between the strings '11101' and '10001' is 2.

Jaccard similarity coefficient is defined as that the proportion of intersection elements of the two sets A and B in the union of A and B. Equation (7) show its formula.

$$J(A, B) = \frac{|A \cap B|}{|A \cup B|} \tag{7}$$

The above eight similarity measures were respectively used under the hood of SMUTE. They are put under test for intellectual inquisitee. SMUTE could evolve to a bagging type of algorithm, sorting out a version that offers the best performance.

Figures 1 and 2 present the changes of average, and standard deviation of Kappa statistics over 100 imbalanced datasets that are pre-processed by the eight versions of SMUTE, respectively. Obviously, all versions of SMUTE are able to improve the performance of imbalanced classification model. The similarity measures by Chebyshev distance, Hamming distance, Jaccard similarity coefficient are the worst methods of the eight. This observation is also confirmed by the following Figures. It can be seen from Fig. 3 that with the increase of under-sampling rate, the average accuracies of the rebalanced classification mode from Chebyshev distance, Hamming distance, Jaccard similarity coefficient versions SMUTE are falling rapidly. The performances of

Minkowski distance SMUTE and Euclidean SMUTE are wound together with a mediocre result. The third best version of SMUTE is obtained by the similarity measure of Manhattan distance. It even could achieve the best performances at certain sampling rates. This phenomenon is reflected in Figs. 1 and 2.

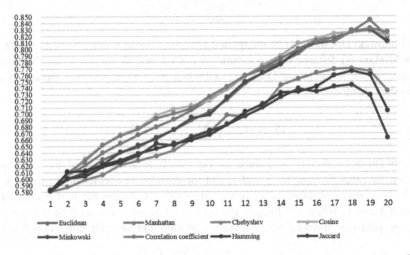

Fig. 1. With the increase of under-sampling rate, the changes of average Kappa statistics of 100 imbalanced dataset of the eight versions of SMUTE.

Figures 1, 2, 3 and 4 shows the performance if these eight methods. The y axes indicates the specific value and the under-sampling rate is in gradually increase from the left to the right side of the x axes.

The similarity measure of Angle Cosine is more suitable for SMUTE than correlation coefficient version. Although, the latter is very close to the former, line of Cosine SMUTE lays above the orange line of correlation coefficient SMUTE at the most of the points. Moreover, Fig. 2 reveals Cosine SMUTE has the minimum standard deviation among the eight versions of SMUTE. Figure 4 displays the changes of reliable accuracy of 1000 imbalanced dataset of the eight versions of SMUTE with the increase of under-sampling rate, reliable accuracy is the product of Kappa and accuracy [15].

In order to more rigorously analyze the distribution variation of processed imbalanced datasets, Figs. 5 and 6 visualize the distribution of an imbalanced dataset after processed by eight versions of SMUTE, respectively. The imbalanced dataset, Ecoli4, is selected from the 100 imbalanced datasets as an example for illustration. The three Figures show the results generated at different under-sampling rates, 0.8, 0.6 and 0.4. K is the top K majority class samples which are similar to all the minority class samples, it is arbitrarily chosen as 12 in this example. In addition, these scatter plots display the first and the second attributes of Ecoli4 to demonstrate the effect of rebalancing the class distribution of the dataset (Fig. 7).

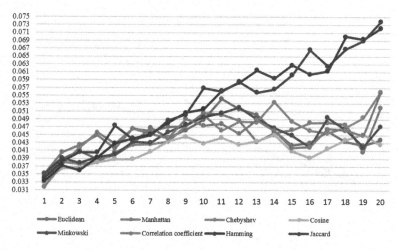

Fig. 2. With the increase of under-sampling rate, the changes of the stand deviation of Kappa statistics of 100 imbalanced dataset of the eight versions of SMUTE.

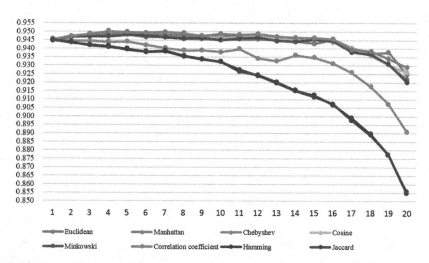

Fig. 3. With the increase of under-sampling rate, the changes of accuracy of 100 imbalanced dataset of the eight versions of SMUTE.

The circles in the scatter plots represent the majority class samples and the asterisks are the minority class samples. The red circles symbolize the selected majority class samples which satisfy K and the under-sampling rate, and the yellow circles are the removed majority class samples. From the results of the experiment where the under-sampling rate in the first scatter plot which is 0.8, it is obviously seen that Angle Cosine similarity version SMUTE is different from the other seven. When the under-sampling rate decreases to 0.6 which is shown in the second scatter plot, comparing with the other seven versions, Angle Cosine similarity version SMUTE has the

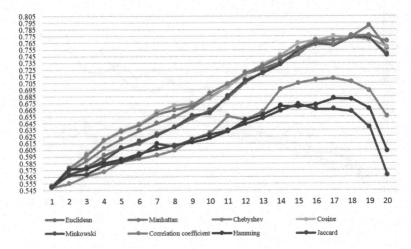

Fig. 4. With the increase of under-sampling rate, the changes of reliable accuracy of 100 imbalanced dataset of the eight versions of SMUTE.

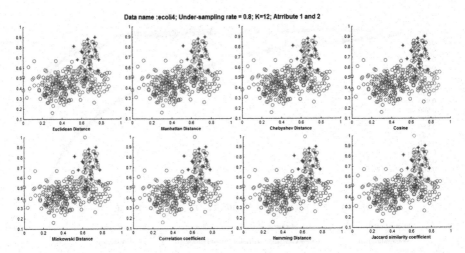

Fig. 5. The distribution of an imbalanced dataset processed by eight versions SMUTE, respectively. Dataset: Ecoli4; under-sampling rate = 0.8; K = 12; Selected Attributes 1 and 2 for display. (Color figure online)

selected samples stand out from most of majority class samples surrounding and they are similar among the minority class samples. However, there is no overlap between majority class samples and minority class samples in this case. Moreover, Cosine similarity version SMUTE is able to completely separate the majority class and minority class in the third scatter plot when the under-sampling rate is at 0.4. The minority class samples are still surrounded by the majority class samples.

In summary, it is possible to implement the most effective version of SMUTE using Angle Cosine after testing out various options. It is apparently useful for easing the

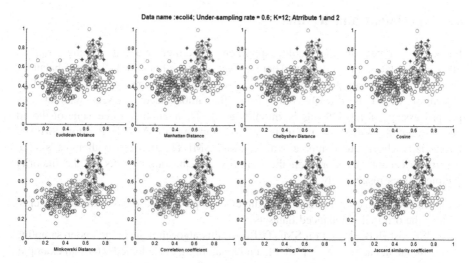

Fig. 6. The distribution of an imbalanced dataset processed by eight versions SMUTE, respectively. Dataset: Ecoli4; under-sampling rate = 0.6; $K = 12$; Selected Attributes 1 and 2 for display. (Color figure online)

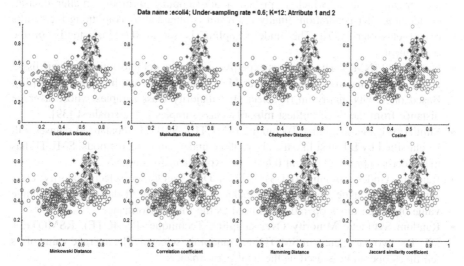

Fig. 7. The distribution of an imbalanced dataset processed by eight versions SMUTE, respectively. Dataset: Ecoli4; under-sampling rate = 0.4; $K = 12$; Selected Attributes 1 and 2 for display. (Color figure online)

imbalanced classification problem. Since it is shown to be superior, Angle Cosine, again will be used in the following experiments for comparing the classification performances of SMUTE and the other existing rebalancing methods.

4 Experiment and Results

For a fair comparison, the whole experiment is verified and tested using stratified 10-cross-validation methodology. There are six other methods for tackling imbalanced classification to be compared with SMUTE. As a baseline reference, the Decision Tree algorithm which is directly built from the imbalanced datasets without any pre-processing of rebalancing is initially used. Then another three state-of-the-art under-sampling methods are put under test to compare with SMUTE. Random SMOTE is specially chosen to compare with random SMUTE, because both have similar constructs except one works on the minority and the other works on the majority sample groups respectively.

- Decision Tree (DT): Decision tree directly adopts tenfold cross-validation to classify the original imbalanced datasets.
- RUSBoost (RB): Based on SMOTEBoost, this algorithm gears the under-sampling technique in each iteration [36], the under-sampling rate adopted 0.6 as an example, which is the same as in the other methods in the experiment.
- BalanceCascade (BC): It adopts the incremental training fashion for boosting. Firstly, it trains a classifier through a training set from under-sampling, and the correctly classified samples are not returned. Then it trains a second classifier from a new training subset, which is picked under-sampling from the smaller dataset. Generation-after generation, finally the result is obtained by combining the results of all classifiers [37]. The under-sampling is set at 0.6 consistently in the experiment.
- NearMiss-2 (NM2): The previous experimental results demonstrate that NearMiss-2 could give competitive results in Near Miss series under-sampling algorithms [38]. Based on K-NN algorithm, it selects the majority class samples whose average distance from the three farthest minority class samples is the smallest [39].
- Similarity Under-sampling Technique (SMUTE): SMUTE rebalanced dataset will be classified by DT, just like the above three under-sampling methods, SMUTE has the under-sampling rate set at 0.6 in the experiment for fairness.
- Random Under-sampling (RUS): RUS + DT. Randomly select majority class samples using under-sampling to rebalance the imbalanced dataset. The average value of its ten times operations is used as the final performance.
- Random Synthetic Minority Over-sampling Technique (RSMOTE): RSMOTE + DT. The two parameters of SMOTE are randomly selected. The average value of its ten times operation is used as the final performance.
- Random SMUTE (RSMUTE): RSMUTE + DT, as RSMOTE, its two parameters are randomly selected and the final performance adopts the average value of its ten times repeated operation.

The mean of the ten times repeated test results is deemed as the final result. The performances of these algorithms are examined using 20 imbalanced datasets from the 100 KEEL datasets, these 20 datasets obtained the worst performance of original imbalanced classification model by decision tree. The characteristics of the 20 selected dataset are described in Table 1. The imbalanced ratios of these datasets between majority class samples and minority class samples ranges from 9.12 to 129.44.

Table 1. Information of testing datasets

Dataset	Attributes	Samples	Majority class	Minority class	Imbalance ratio (maj/min)
abalone-17_vs_7-8-9-10	8	2338	2280	58	39.31
abalone-19_vs_10-11-12-13	8	1622	1590	32	49.69
abalone-20_vs_8-9-10	8	1916	1890	26	72.69
abalone-21_vs_8	8	581	567	14	40.50
abalone19	8	4174	4142	32	129.44
abalone9-18	8	731	689	42	16.40
cleveland-0_vs_4	13	177	164	13	12.62
glass-0-1-4-6_vs_2	9	205	188	17	11.06
glass-0-1-5_vs_2	9	172	155	17	9.12
glass-0-1-6_vs_2	9	192	175	17	10.29
glass2	9	214	197	17	11.59
poker-8-9_vs_6	10	1485	1460	25	58.40
poker-8_vs_6	10	1477	1460	17	85.88
poker-9_vs_7	10	244	236	8	29.50
winequality-red-3_vs_5	11	691	681	10	68.10
winequality-red-8_vs_6-7	11	855	837	18	46.50
winequality-red-8_vs_6	11	656	638	18	35.44
winequality-white-3-9_vs_5	11	1482	1457	25	58.28
winequality-white-9_vs_4	11	168	163	5	32.60
yeast-0-5-6-7-9_vs_4	8	528	477	51	9.35

For the experimentation, the simulation software is programmed by in MATLAB version 2014b. The simulation computing platform is CPU: E5-1650 V2 @ 3.50 GHz, RAM: 62 GB.

In the experimentation, Kappa statistics is used to supplement the accuracy measure for evaluating the goodness of the classification model. They are recorded in the Tables 1 and 2. Moreover, the performances in terms of BER, MCC, Precision, Recall and F1 measure are respectively tabulated in the Tables 4, 5, 6, 7 and 8 in Appendix.

The first columns of Tables 2 and 3 confirm that the imbalanced classification problem exists where the accuracy rate seems high but with very low credibility (Kappa). There are results by the four under-sampling methods with the same under-sampling rate listed in the second part of these two Tables. RB obtained worse performances in both of Kappa and accuracy when upon the difficulty in processed the imbalanced datasets. Essentially, BC is the ensemble type of under-sampling methods, they blend AdaBoosting methods for easing the imbalanced data proportion. However, it can be observed that the performance of SMUTE is clearly better than BC when they are at the same sampling rate. Furthermore, with the increase of Kappa statistics,

Table 2. Kappa statistics of each datasets with different methods

Kappa	DT	RB (0.6)	NMS2 (0.6)	BC (0.6)	SMUTE (0.6)	RUS	RSMOTE	RSMUTE
abalone-17_vs_7-8-9-10	0.139	0.157	0.069	0.164	0.164	0.190	0.200	0.190
abalone-19_vs_10-11-12-13	0.000	−0.016	0.007	0.032	0.021	0.046	0.008	0.027
abalone-20_vs_8-9-10	0.097	0.080	0.042	0.195	0.072	0.145	0.150	0.147
abalone-21_vs_8	0.208	0.235	0.189	0.300	0.271	0.369	0.280	0.351
abalone19	−0.007	0.020	0.012	0.019	0.017	0.032	0.021	0.021
abalone9-18	0.174	0.192	0.116	0.222	0.177	0.164	0.250	0.269
cleveland-0_vs_4	0.210	0.613	0.209	0.367	0.548	0.351	0.296	0.427
glass-0-1-4-6_vs_2	−0.006	0.131	0.043	0.062	0.218	0.111	0.119	0.201
glass-0-1-5_vs_2	−0.008	0.116	0.151	0.116	0.203	0.116	0.132	0.167
glass-0-1-6_vs_2	0.027	−0.007	0.135	0.156	0.303	0.074	0.138	0.147
glass2	−0.013	0.113	0.097	0.163	0.171	0.100	0.140	0.104
poker-8-9_vs_6	−0.007	0.243	0.104	0.155	0.080	0.131	0.346	0.139
poker-8_vs_6	0.055	0.039	0.061	0.248	0.087	0.099	0.364	0.161
poker-9_vs_7	−0.009	0.095	0.063	0.042	0.254	0.198	0.150	0.220
winequality-red-3_vs_5	−0.010	0.039	0.021	0.115	0.060	0.072	0.084	0.088
winequality-red-8_vs_6-7	0.000	0.049	0.060	0.054	0.164	0.060	0.076	0.064
winequality-red-8_vs_6	0.021	0.137	0.077	0.154	0.142	0.157	0.125	0.132
winequality-white-3-9_vs_5	0.034	0.095	0.028	0.054	0.181	0.095	0.138	0.150
winequality-white-9_vs_4	−0.010	0.233	0.081	0.000	0.465	0.123	0.149	0.114
yeast-0-5-6-7-9_vs_4	0.287	0.341	0.268	0.358	0.383	0.271	0.274	0.394
Average	0.059	0.145	0.092	0.149	0.199	0.145	0.172	0.176
Stand deviation	0.091	0.140	0.068	0.106	0.138	0.090	0.097	0.108

SMUTE is able to efficiently retain a high accuracy than the other three methods, besides RB.

RSMOTE and RSMUTE respectively stand for over-sampling and under-sampling techniques. Quite evidently, these two methods are better than the random under-sampling scheme. In comparison, the results of RSMOTE and RSMUTE are close, but RSMUTE could achieve higher correctness from the rebalanced classification model. Figure 8 shows that the total average values of Kappa statistics and the other auxiliary evaluation metrics for imbalanced classification models. The average values of Kappa statistics and accuracy of these 20 datasets are shown at the bottoms of the Tables SMUTE and RSMUTE obtained the best performances in their own groups considering the priority of Kappa (which means how well the model can generalized) over accuracy. As it can be seen from RSMUTE that the parameters of SMUTE are crucial to the good performance of rebalanced classification. Choosing the right set of parameter values is also applicable for SMOTE. In general, when putting RSMOTE and RSMUTE vis-à-vis, the results show that SMUTE could score more wins over the 20 datasets, and better average accuracy and Kappa.

Table 3. Accuracy of each datasets with different methods.

Accuracy	DT	RB (0.6)	NMS2 (0.6)	BC (0.6)	SMUTE (0.6)	RUS	RSMOTE	RSMUTE
abalone-17_vs_7-8-9-10	0.960	0.943	0.715	0.829	0.847	0.930	0.932	0.928
abalone-19_vs_10-11-12-13	0.980	0.868	0.613	0.678	0.782	0.921	0.917	0.947
abalone-20_vs_8-9-10	0.976	0.925	0.798	0.946	0.821	0.963	0.939	0.953
abalone-21_vs_8	0.973	0.969	0.935	0.971	0.895	0.953	0.947	0.963
abalone19	0.985	0.891	0.752	0.713	0.725	0.964	0.938	0.946
abalone9-18	0.917	0.871	0.718	0.784	0.770	0.870	0.877	0.892
cleveland-0_vs_4	0.904	0.949	0.829	0.932	0.949	0.831	0.902	0.924
glass-0-1-4-6_vs_2	0.902	0.854	0.590	0.624	0.786	0.803	0.849	0.877
glass-0-1-5_vs_2	0.825	0.884	0.670	0.628	0.819	0.738	0.800	0.848
glass-0-1-6_vs_2	0.834	0.807	0.648	0.672	0.776	0.686	0.816	0.860
glass2	0.869	0.846	0.618	0.720	0.760	0.756	0.827	0.863
poker-8-9_vs_6	0.979	0.969	0.885	0.925	0.934	0.952	0.977	0.966
poker-8_vs_6	0.976	0.975	0.903	0.908	0.905	0.951	0.983	0.978
poker-9_vs_7	0.959	0.943	0.824	0.770	0.909	0.920	0.961	0.959
winequality-red-3_vs_5	0.970	0.954	0.509	0.980	0.913	0.947	0.956	0.970
winequality-red-8_vs_6-7	0.979	0.943	0.930	0.765	0.979	0.938	0.919	0.955
winequality-red-8_vs_6	0.954	0.939	0.907	0.896	0.970	0.895	0.926	0.948
winequality-white-3-9_vs_5	0.968	0.920	0.717	0.736	0.970	0.945	0.952	0.972
winequality-white-9_vs_4	0.964	0.964	0.941	0.030	0.953	0.945	0.958	0.968
yeast-0-5-6-7-9_vs_4	0.877	0.862	0.788	0.816	0.829	0.829	0.848	0.871
Average	0.938	0.914	0.764	0.766	0.865	0.887	0.911	0.929
Stand deviation	0.050	0.048	0.127	0.203	0.080	0.082	0.054	0.043

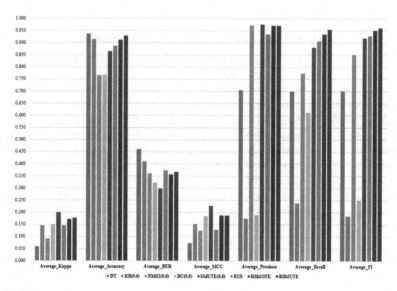

Fig. 8. The total average values of each auxiliary evaluation metrics of each algorithm for all datasets.

5   Conclusion

In this paper, we proposed a novel class rebalancing method for subsiding the imbalanced classification problems, called Similarity Majority Under-sampling Technique, SMUTE. It removes a certain amount of majority class samples through sorting the similarity between each majority class sample and minority class samples; the top similar majority class samples are retained. The experimental results show that SMUTE could exceed the performance of the state-of-the-arts using selective down-sampling method under the control circumstances. SMUTE is believed to be useful in data mining, particularly as a pre-processing approach for fixing training dataset that contains only few but important minority class samples. It works by preserving all the available minority class samples (which are supposed to be precious), and reducing the overly large population of majority class samples, keeping only those majority class samples that are similar to the rare samples by data distances.

Acknowledgement. The authors are thankful to the financial support from the research grant, #MYRG2016-00069, titled 'Nature-Inspired Computing and Metaheuristics Algorithms for Optimizing Data Mining Performance' offered by RDAO/FST, University of Macau and Macau SAR government.

Appendix

See Tables 4, 5, 6, 7 and 8.

Table 4. BER of each datasets with different methods.

BER	DT	RB (0.6)	NMS2 (0.6)	BC (0.6)	SMUTE (0.6)	RUS	RSMOTE	RSMUTE
abalone-17_vs_7-8-9-10	0.430	0.390	0.287	0.155	0.285	0.358	0.322	0.341
abalone-19_vs_10-11-12-13	0.500	0.527	0.475	0.364	0.343	0.445	0.481	0.485
abalone-20_vs_8-9-10	0.423	0.379	0.324	0.255	0.168	0.379	0.300	0.362
abalone-21_vs_8	0.404	0.399	0.277	0.289	0.105	0.268	0.271	0.234
abalone19	0.504	0.427	0.415	0.315	0.299	0.465	0.461	0.459
abalone9-18	0.410	0.359	0.328	0.215	0.209	0.384	0.311	0.311
cleveland-0_vs_4	0.406	0.205	0.370	0.355	0.234	0.272	0.325	0.244
glass-0-1-4-6_vs_2	0.401	0.428	0.428	0.419	0.238	0.417	0.409	0.389
glass-0-1-5_vs_2	0.499	0.457	0.335	0.364	0.368	0.415	0.400	0.392
glass-0-1-6_vs_2	0.498	0.504	0.352	0.313	0.253	0.432	0.403	0.406
glass2	0.505	0.433	0.415	0.314	0.190	0.413	0.400	0.435
poker-8-9_vs_6	0.502	0.350	0.411	0.290	0.367	0.403	0.310	0.414
poker-8_vs_6	0.482	0.478	0.370	0.285	0.402	0.445	0.295	0.406

(*continued*)

Table 4. (*continued*)

BER	DT	RB (0.6)	NMS2 (0.6)	BC (0.6)	SMUTE (0.6)	RUS	RSMOTE	RSMUTE
poker-9_vs_7	0.404	0.452	0.283	0.421	0.320	0.253	0.308	0.270
winequality-red-3_vs_5	0.508	0.467	0.298	0.454	0.397	0.401	0.426	0.429
winequality-red-8_vs_6-7	0.500	0.464	0.452	0.338	0.426	0.428	0.433	0.454
winequality-red-8_vs_6	0.485	0.409	0.437	0.323	0.428	0.379	0.407	0.416
winequality-white-3-9_vs_5	0.467	0.375	0.480	0.272	0.373	0.425	0.368	0.425
winequality-white-9_vs_4	0.503	0.406	0.219	0.500	0.436	0.161	0.144	0.186
yeast-0-5-6-7-9_vs_4	0.366	0.295	0.274	0.207	0.132	0.333	0.329	0.271
Average	0.460	0.410	0.361	0.322	0.299	0.374	0.355	0.366
Stand deviation	0.047	0.072	0.073	0.083	0.100	0.077	0.077	0.083

Table 5. MCC of each datasets with different methods.

MCC	DT	RB (0.6)	NMS2 (0.6)	BC (0.6)	SMUTE (0.6)	RUS	RSMOTE	RSMUTE
abalone-17_vs_7-8-9-10	0.141	0.165	0.146	0.275	0.186	0.206	0.226	0.216
abalone-19_vs_10-11-12-13	0.000	−0.023	0.016	0.081	0.044	0.057	0.014	0.030
abalone-20_vs_8-9-10	0.104	0.110	0.099	0.248	0.144	0.149	0.195	0.174
abalone-21_vs_8	0.212	0.238	0.236	0.211	0.334	0.347	0.312	0.381
abalone19	−0.007	0.042	0.035	0.071	0.053	0.039	0.031	0.032
abalone9-18	0.179	0.206	0.175	0.306	0.236	0.178	0.272	0.289
cleveland-0_vs_4	0.213	0.613	0.230	0.387	0.566	0.219	0.310	0.445
glass-0-1-4-6_vs_2	0.242	0.132	0.072	0.092	0.229	0.097	0.133	0.209
glass-0-1-5_vs_2	−0.006	0.129	0.201	0.166	0.226	0.088	0.152	0.179
glass-0-1-6_vs_2	0.030	−0.007	0.184	0.221	0.364	0.074	0.152	0.160
glass2	−0.014	0.115	0.116	0.219	0.174	0.098	0.151	0.111
poker-8-9_vs_6	−0.008	0.248	0.116	0.247	0.096	0.145	0.357	0.148
poker-8_vs_6	0.059	0.039	0.098	0.129	0.097	0.049	0.381	0.167
poker-9_vs_7	−0.009	0.095	0.103	0.068	0.287	0.040	0.155	0.226
winequality-red-3_vs_5	−0.011	0.043	0.097	0.119	0.071	0.039	0.095	0.097
winequality-red-8_vs_6-7	0.000	0.052	0.066	0.110	0.168	0.072	0.088	0.069
winequality-red-8_vs_6	0.022	0.142	0.085	0.192	0.142	0.159	0.135	0.140
winequality-white-3-9_vs_5	0.039	0.124	0.031	0.132	0.186	0.107	0.159	0.156
winequality-white-9_vs_4	−0.014	0.241	0.077	0.000	0.519	0.106	0.161	0.117
yeast-0-5-6-7-9_vs_4	0.295	0.348	0.313	0.409	0.436	0.285	0.286	0.406
Average	0.073	0.153	0.125	0.184	0.228	0.128	0.188	0.188
Stand deviation	0.100	0.140	0.075	0.105	0.145	0.083	0.100	0.112

Table 6. Precision of each datasets with different methods.

Precision	DT	RB (0.6)	NMS2 (0.6)	BC (0.6)	SMUTE (0.6)	RUS	RSMOTE	RSMUTE
abalone-17_vs_7-8-9-10	0.979	0.143	0.990	0.113	0.987	0.983	0.984	0.984
abalone-19_vs_10-11-12-13	0.000	0.011	0.982	0.036	0.984	0.983	0.981	0.981
abalone-20_vs_8-9-10	0.988	0.060	0.992	0.132	0.994	0.990	0.992	0.990
abalone-21_vs_8	0.981	0.300	0.985	0.429	0.991	0.986	0.988	0.989
abalone19	0.992	0.018	0.994	0.017	0.995	0.993	0.993	0.993
abalone9-18	0.953	0.190	0.969	0.181	0.973	0.957	0.966	0.966
cleveland-0_vs_4	0.300	0.667	0.954	0.571	0.965	0.684	0.954	0.966
glass-0-1-4-6_vs_2	0.364	0.190	0.932	0.115	0.937	0.932	0.933	0.936
glass-0-1-5_vs_2	0.903	0.286	0.940	0.159	0.932	0.837	0.923	0.926
glass-0-1-6_vs_2	0.914	0.083	0.950	0.171	0.968	0.913	0.930	0.929
glass2	0.921	0.167	0.942	0.169	0.937	0.837	0.938	0.931
poker-8-9_vs_6	0.000	0.216	0.986	0.621	0.985	0.987	0.989	0.986
poker-8_vs_6	0.989	0.045	0.992	0.297	0.990	0.990	0.993	0.991
poker-9_vs_7	0.967	0.125	0.980	0.056	0.986	0.883	0.975	0.978
winequality-red-3_vs_5	0.985	0.042	0.998	0.167	0.987	0.989	0.988	0.988
winequality-red-8_vs_6-7	0.000	0.057	0.981	0.049	0.981	0.981	0.982	0.981
winequality-red-8_vs_6	0.974	0.133	0.977	0.121	0.975	0.882	0.978	0.977
winequality-white-3-9_vs_5	0.984	0.073	0.984	0.045	0.987	0.986	0.988	0.986
winequality-white-9_vs_4	0.000	0.333	0.974	0.030	0.994	0.982	0.984	0.978
yeast-0-5-6-7-9_vs_4	0.929	0.351	0.956	0.315	0.974	0.939	0.939	0.950
Average	0.706	0.175	0.973	0.190	0.976	0.936	0.970	0.970
Stand deviation	0.400	0.153	0.020	0.170	0.019	0.077	0.024	0.022

Table 7. Recall of each datasets with different methods.

Recall	DT	RB (0.6)	NMS2 (0.6)	BC (0.6)	SMUTE (0.6)	RUS	RSMOTE	RSMUTE
abalone-17_vs_7-8-9-10	0.980	0.259	0.715	0.862	0.889	0.945	0.945	0.942
abalone-19_vs_10-11-12-13	0.000	0.063	0.617	0.594	0.791	0.936	0.934	0.965
abalone-20_vs_8-9-10	0.987	0.308	0.802	0.538	0.824	0.973	0.946	0.962
abalone-21_vs_8	0.991	0.214	0.947	0.643	0.901	0.965	0.957	0.972
abalone19	0.993	0.250	0.754	0.656	0.726	0.970	0.945	0.952
abalone9-18	0.959	0.381	0.724	0.786	0.778	0.903	0.901	0.918
cleveland-0_vs_4	0.231	0.615	0.860	0.308	0.982	0.847	0.941	0.952
glass-0-1-4-6_vs_2	0.235	0.235	0.595	0.529	0.819	0.846	0.902	0.931
glass-0-1-5_vs_2	0.901	0.118	0.680	0.647	0.864	0.771	0.851	0.906
glass-0-1-6_vs_2	0.904	0.118	0.646	0.706	0.785	0.716	0.864	0.918
glass2	0.939	0.235	0.620	0.647	0.791	0.793	0.870	0.920
poker-8-9_vs_6	0.000	0.320	0.895	0.720	0.947	0.964	0.987	0.980
poker-8_vs_6	0.987	0.059	0.910	0.647	0.914	0.960	0.990	0.988

(continued)

Table 7. (*continued*)

Recall	DT	RB (0.6)	NMS2 (0.6)	BC (0.6)	SMUTE (0.6)	RUS	RSMOTE	RSMUTE
poker-9_vs_7	0.991	0.125	0.834	0.375	0.918	0.935	0.985	0.980
winequality-red-3_vs_5	0.984	0.100	0.503	0.100	0.924	0.958	0.967	0.982
winequality-red-8_vs_6-7	0.000	0.111	0.946	0.556	0.998	0.954	0.934	0.973
winequality-red-8_vs_6	0.980	0.222	0.926	0.444	0.994	0.911	0.945	0.969
winequality-white-3-9_vs_5	0.983	0.320	0.724	0.720	0.982	0.957	0.964	0.986
winequality-white-9_vs_4	0.000	0.200	0.963	1.000	0.958	0.959	0.971	0.989
yeast-0-5-6-7-9_vs_4	0.935	0.510	0.803	0.765	0.834	0.868	0.891	0.905
Average	0.699	0.238	0.773	0.612	0.881	0.907	0.934	0.954
Stand deviation	0.412	0.141	0.133	0.195	0.081	0.073	0.041	0.028

Table 8. F1-measure of each datasets with different methods.

F1	DT	RB (0.6)	NMS2 (0.6)	BC (0.6)	SMUTE (0.6)	RUS	RSMOTE	RSMUTE
abalone-17_vs_7-8-9-10	0.979	0.184	0.828	0.200	0.895	0.962	0.964	0.962
abalone-19_vs_10-11-12-13	0.000	0.018	0.755	0.068	0.876	0.958	0.956	0.973
abalone-20_vs_8-9-10	0.988	0.101	0.884	0.212	0.900	0.981	0.968	0.976
abalone-21_vs_8	0.986	0.250	0.965	0.514	0.943	0.975	0.972	0.981
abalone19	0.993	0.034	0.856	0.034	0.837	0.981	0.968	0.972
abalone9-18	0.956	0.254	0.828	0.295	0.863	0.928	0.932	0.941
cleveland-0_vs_4	0.261	0.640	0.900	0.400	0.973	0.863	0.946	0.958
glass-0-1-4-6_vs_2	0.286	0.211	0.720	0.189	0.871	0.879	0.915	0.932
glass-0-1-5_vs_2	0.900	0.167	0.778	0.256	0.889	0.818	0.882	0.914
glass-0-1-6_vs_2	0.907	0.098	0.756	0.276	0.859	0.778	0.893	0.922
glass2	0.929	0.195	0.742	0.268	0.849	0.837	0.900	0.924
poker-8-9_vs_6	0.000	0.258	0.937	0.837	0.965	0.975	0.988	0.983
poker-8_vs_6	0.988	0.051	0.948	0.407	0.949	0.974	0.992	0.989
poker-9_vs_7	0.979	0.125	0.895	0.097	0.949	0.946	0.980	0.979
winequality-red-3_vs_5	0.984	0.059	0.663	0.125	0.951	0.972	0.977	0.984
winequality-red-8_vs_6-7	0.000	0.075	0.963	0.090	0.989	0.967	0.957	0.977
winequality-red-8_vs_6	0.976	0.167	0.951	0.190	0.984	0.933	0.961	0.973
winequality-white-3-9_vs_5	0.983	0.119	0.823	0.084	0.984	0.971	0.975	0.986
winequality-white-9_vs_4	0.000	0.250	0.967	0.058	0.975	0.968	0.977	0.983
yeast-0-5-6-7-9_vs_4	0.931	0.416	0.872	0.446	0.898	0.898	0.913	0.926
Average	0.701	0.184	0.852	0.252	0.920	0.928	0.951	0.962
Stand deviation	0.407	0.141	0.090	0.190	0.050	0.060	0.032	0.025

References

1. Weiss, G.M., Provost, F.: Learning when training data are costly: the effect of class distribution on tree induction. J. Artif. Intell. Res. **19**, 315–354 (2003)
2. Li, J., Fong, S., Sung, Y., Cho, K., Wong, R., Wong, K.K.: Adaptive swarm cluster-based dynamic multi-objective synthetic minority oversampling technique algorithm for tackling binary imbalanced datasets in biomedical data classification. BioData Min. **9**(1), 37 (2016)
3. Cao, H., Li, X.-L., Woon, D.Y.-K., Ng, S.-K.: Integrated oversampling for imbalanced time series classification. IEEE Trans. Knowl. Data Eng. **25**(12), 2809–2822 (2013)
4. Kubat, M., Holte, R.C., Matwin, S.: Machine learning for the detection of oil spills in satellite radar images. Mach. Learn. **30**(2–3), 195–215 (1998)
5. Li, J., Fong, S., Mohammed, S., Fiaidhi, J.: Improving the classification performance of biological imbalanced datasets by swarm optimization algorithms. J. Supercomput. **72**(10), 3708–3728 (2016)
6. Chawla, N.V.: C4. 5 and imbalanced data sets: investigating the effect of sampling method, probabilistic estimate, and decision tree structure (2002)
7. Tang, Y., Zhang, Y.-Q., Chawla, N.V., Krasser, S.: SVMs modeling for highly imbalanced classification. IEEE Trans. Syst. Man Cybern. Part B (Cybern.) **39**(1), 281–288 (2009)
8. Li, J., Fong, S., Yuan, M., Wong, R.K.: Adaptive multi-objective swarm crossover optimization for imbalanced data classification. In: Li, J., Li, X., Wang, S., Li, J., Sheng, Q. Z. (eds.) ADMA 2016. LNCS (LNAI), vol. 10086, pp. 374–390. Springer, Cham (2016). https://doi.org/10.1007/978-3-319-49586-6_25
9. Stone, E.A.: Predictor performance with stratified data and imbalanced classes. Nat. Methods **11**(8), 782 (2014)
10. Guo, H., Viktor, H.L.: Learning from imbalanced data sets with boosting and data generation: the databoost-im approach. ACM SIGKDD Explor. Newsl. **6**(1), 30–39 (2004)
11. Weiss, G.M.: Learning with rare cases and small disjuncts. In: ICML, pp. 558–565 (1995)
12. Weiss, G.M.: Mining with rarity: a unifying framework. ACM SIGKDD Explor. Newsl. **6** (1), 7–19 (2004)
13. Arunasalam, B., Chawla, S.: CCCS: a top-down associative classifier for imbalanced class distribution. In: Proceedings of the 12th ACM SIGKDD International Conference on Knowledge Discovery and Data Mining, pp. 517–522. ACM (2006)
14. Breiman, L., Friedman, J., Stone, C.J., Olshen, R.A.: Classification and Regression Trees. CRC Press, Boca Raton (1984)
15. Li, J., Fong, S., Wong, R.K., Chu, V.W.: Adaptive multi-objective swarm fusion for imbalanced data classification. Inf. Fusion **39**, 1–24 (2018)
16. Drummond, C., Holte, R.C.: C4. 5, class imbalance, and cost sensitivity: why under-sampling beats over-sampling. In: Workshop on Learning from Imbalanced Datasets II, Citeseer (2003)
17. Chawla, N.V., Bowyer, K.W., Hall, L.O., Kegelmeyer, W.P.: SMOTE: synthetic minority over-sampling technique. J. Artif. Intell. Res. **16**, 321–357 (2002)
18. Han, H., Wang, W.-Y., Mao, B.-H.: Borderline-SMOTE: a new over-sampling method in imbalanced data sets learning. In: Huang, D.-S., Zhang, X.-P., Huang, G.-B. (eds.) ICIC 2005. LNCS, vol. 3644, pp. 878–887. Springer, Heidelberg (2005). https://doi.org/10.1007/11538059_91
19. Hu, S., Liang, Y., Ma, L., He, Y.: MSMOTE: improving classification performance when training data is imbalanced. In: Second International Workshop on Computer Science and Engineering, WCSE 2009, pp. 13–17. IEEE (2009)

20. Kubat, M., Matwin, S.: Addressing the curse of imbalanced training sets: one-sided selection. In: ICML, pp. 179–186 (1997)
21. Chen, X., Gerlach, B., Casasent, D.: Pruning support vectors for imbalanced data classification. In: IJCNN 2005, Proceedings, pp. 1883–1888. IEEE (2005)
22. Raskutti, B., Kowalczyk, A.: Extreme re-balancing for SVMs: a case study. ACM SIGKDD Explor. Newsl. 6(1), 60–69 (2004)
23. Estabrooks, A., Japkowicz, N.: A mixture-of-experts framework for learning from imbalanced data sets. In: Hoffmann, F., Hand, D.J., Adams, N., Fisher, D., Guimaraes, G. (eds.) IDA 2001. LNCS, vol. 2189, pp. 34–43. Springer, Heidelberg (2001). https://doi.org/10.1007/3-540-44816-0_4
24. Japkowicz, N., Stephen, S.: The class imbalance problem: a systematic study. Intell. Data Anal. 6(5), 429–449 (2002)
25. Quinlan, J.R.: Bagging, boosting, and C4. 5. In: AAAI/IAAI, vol. 1, pp. 725–730 (1996)
26. Sun, Y., Kamel, M.S., Wang, Y.: Boosting for learning multiple classes with imbalanced class distribution. In: Sixth International Conference on ICDM 2006, pp. 592–602. IEEE (2006)
27. Alcalá, J., Fernández, A., Luengo, J., Derrac, J., García, S., Sánchez, L., Herrera, F.: Keel data-mining software tool: data set repository, integration of algorithms and experimental analysis framework. J. Multiple-Valued Logic Soft Comput. 17(2–3), 255–287 (2010)
28. Li, J., Fong, S., Zhuang, Y.: Optimizing SMOTE by metaheuristics with neural network and decision tree. In: 3rd International Symposium on Computational and Business Intelligence (ISCBI), pp. 26–32. IEEE (2015)
29. Viera, A.J., Garrett, J.M.: Understanding interobserver agreement: the kappa statistic. Fam. Med. 37(5), 360–363 (2005)
30. Cha, S.-H.: Comprehensive survey on distance/similarity measures between probability density functions. City 1(2), 1 (2007)
31. Nguyen, H.V., Bai, L.: Cosine similarity metric learning for face verification. In: Kimmel, R., Klette, R., Sugimoto, A. (eds.) ACCV 2010. LNCS, vol. 6493, pp. 709–720. Springer, Heidelberg (2011). https://doi.org/10.1007/978-3-642-19309-5_55
32. Santini, S., Jain, R.: Similarity measures. IEEE Trans. Pattern Anal. Mach. Intell. 21(9), 871–883 (1999)
33. Ahlgren, P., Jarneving, B., Rousseau, R.: Requirements for a cocitation similarity measure, with special reference to Pearson's correlation coefficient. J. Am. Soc. Inform. Sci. Technol. 54(6), 550–560 (2003)
34. Xu, Z., Xia, M.: Distance and similarity measures for hesitant fuzzy sets. Inf. Sci. 181(11), 2128–2138 (2011)
35. Choi, S.-S., Cha, S.-H., Tappert, C.C.: A survey of binary similarity and distance measures. J. Syst. Cybern. Inform. 8(1), 43–48 (2010)
36. Seiffert, C., Khoshgoftaar, T.M., Van Hulse, J., Napolitano, A.: RUSBoost: a hybrid approach to alleviating class imbalance. IEEE Trans. Syst. Man Cybern.-Part A: Syst. Hum. 40(1), 185–197 (2010)
37. Liu, X.-Y., Wu, J., Zhou, Z.-H.: Exploratory undersampling for class-imbalance learning. IEEE Trans. Syst. Man Cybern. Part B (Cybern.) 39(2), 539–550 (2009)
38. Mani, I., Zhang, I.: kNN approach to unbalanced data distributions: a case study involving information extraction. In: Proceedings of Workshop on Learning from Imbalanced Datasets (2003)
39. He, H., Garcia, E.A.: Learning from imbalanced data. IEEE Trans. Knowl. Data Eng. 21(9), 1263–1284 (2009)

Rank Forest: Systematic Attribute
Sub-spacing in Decision Forest

Zaheer Babar[1(✉)], Md Zahidul Islam[2(✉)], and Sameen Mansha[3]

[1] Institute of Computing and Information Sciences, Radboud University,
Nijmegen, The Netherlands
z.babar@cs.ru.nl
[2] School of Computing and Mathematics, Charles Sturt University, Bathurst, Australia
zislam@csu.edu.au
[3] School of Information Technology and Electrical Engineering, University of Queensland,
Brisbane, Australia
s.mansha@uqconnect.edu.au

Abstract. Decision Trees are well known classification algorithms that are also appreciated for their capacity for knowledge discovery. In the literature two major shortcomings of decision trees have been pointed out: (1) instability, and (2) high computational cost. These problems have been addressed to some extent through ensemble learning techniques such as Random Forest. Unlike decision trees where the whole attribute space of a dataset is used to discover the best test attribute for a node, in Random Forest a random subspace of attributes is first selected from which the test attribute for a node is then identified. The property that randomly selects an attribute subspace can cause the selection of all/many poor quality attributes in a subspace resulting in an individual tree with low accuracy. Therefore, in this paper we propose a probabilistic selection of attributes (instead of a random selection) where the probability of the selection of an attribute is proportionate to its quality. Although we developed this approach independently, after the research was completed we discovered that some existing techniques also took the same approach. While in this paper we use mutual information as a measure of an attribute quality, the papers in the literature used information gain ratio and a t-test as the measure. The proposed technique has been evaluated using nine different datasets and a stable performance can be seen in terms of the accuracy (ensemble accuracy and individual tree accuracy) and efficiency.

Keywords: Ensemble learning · Decision forest · Reduced attribute subspace
Random Forest

1 Introduction

With the growth in data generation, the effective arrangement and recovery of massive data has become progressively troublesome. The prime purpose behind the systematic arrangement, is to discover knowledge from generated data for future planning and directives. Massive research has been conducted and also in progress to devise more

© Springer Nature Singapore Pte Ltd. 2018
Y. L. Boo et al. (Eds.): AusDM 2017, CCIS 845, pp. 24–37, 2018.
https://doi.org/10.1007/978-981-13-0292-3_2

promising solutions for efficient data arrangement. Till now, significant advancements have been made in the field of data mining and machine learning that provide promising ways to process such large datasets [1]. Specifically, two learning tasks (classification and clustering) have become dominant ways to perform knowledge discovery and pattern mining. In clustering, the objective is to divide the data into groups (clusters) on the basis of calculated similarity or minimal distance among data instances. While classification refers to the assignment of a category (from previously defined categories) to newly generated instances, on the basis of identified patterns in previously categorized data.

In data mining operations, we use to represent a dataset as a two dimensional array, as a combination of m data instances/records/rows $D = \{R_1, R_2, R_3, \dots, R_m\}$. It can also be represented as a combination of n attributes/features/columns $D = \{A_1, A_2, A_3, \dots, A_n, C\}$. In normal classification operations, a data instance contains values from n non-class attributes that provide a base to get a value for class attribute C (values from predefined classes/categories), where $C = \{c_1, c_2, c_3, \dots\}$. Each data instance gets a category label (class label) from predefined categories. There are two types of datasets that have been defined to perform classification tasks (1) Existing dataset (termed as Training data) contains value of class attribute (class label) for each data instance (2) Unseen dataset (termed as Testing data) does not contain value for class attribute. In classification task, we build a classifier on training data (which contains predefined classes) and predict the class values for testing data.

In general, classifiers can be divided into two categories: (1) stable classifiers (2) unstable classifiers. Stable classifiers (e.g., SVM, k- nearest neighbors etc.) possess the ability of generalization and are not significantly influenced by a slight change in training dataset. While in case of unstable classifiers (Decision Tree, Neural Network, Linear Regression etc.) a slight change in training dataset might change the learned classifier performance.

In this paper, we focus on decision trees. Despite the popularity of decision trees, they are known to be unstable. Computational complexity is also one of the major overhead of decision tree [2] since building an optimal decision tree is NP hard [3]. In a decision tree, to nominate a node as a splitting point, we need to iterate through all attributes (to find their gain score) and select best attribute with maximum gain score ($O(n)$). After that, in order to find the best value for splitting we have to find entropy score against each unique value of the selected attribute. The whole process need to be followed to select a single splitting node. This characteristic of decision tree marks it more expensive especially in case we have data with higher dimensions.

In order to overcome the instability of classifiers, various solutions have been suggested in the literature, in particular ensemble learning. Ensemble learning is one of the techniques which trains a set of unstable or weak classifiers on different samples from original training data. It classifies the test dataset by taking aggregate (vote) of predictions from multiple classifiers for each test data instance [4]. Numerous ensemble based techniques have been discussed in literature including Bagging, Random Subspace, Boosting, and Random Forest.

Random Forest [5] is one of the decision forest algorithms that is a fusion of Bagging and Random Subspace. Random Forest slightly deals with potential problems related to decision tree. It extracts K random samples from original training data and learns K tree predictors on each of the sample datasets. A tree predictor in Random Forest is slightly different as compare to the decision tree algorithm. Decision tree splits data on the base of an attribute with highest information gain score among all attributes of the dataset. While in case of Random Forest, tree predictor does not consider the whole set of attributes. Rather, it takes a random sample from the set of attributes and applies information gain based ranking over that sample. The randomness of Random Forest ensures that each of the tree predictors must demonstrate slightly different construction as compared to others, which ultimately makes a forest diverse in nature. Furthermore, randomness property by some means also addresses the problem related to computational complexity of the construction of decision tree [6] (time complexity to build single tree $O(n\,m\,\log m)$). Nevertheless, the property of Random Forest for attributes sampling may lead to another problem. Suppose A_x is a subset of attributes of D, that possess more discriminative score than another subset A_y and s denotes sampled attribute subset from A, where $\{A_x, A_y\} \subset A$ or $A = A_x \cup A_y$, and $|A_x| \ll |A_y|$. Let's consider a case where a small amount of the attributes in s belong to A_x, while majority belongs to A_y. A worst case may occur where none of the sampled attributes belongs to set A_x. In this case stability and prediction capability of the model must be reconsidered.

This paper addresses both the above stated problems related to decision trees. Especially it focuses on addressing the computationally expensive construction of decision trees and their instability. We propose a modification of Random Forest that first computes the goodness (i.e. the capacity to discriminate among different classes) of attributes and then uses the goodness values to probabilistically select a subspace of attributes where the probability of the selection of an attribute is proportionate to its goodness value. While we developed this approach independently, after the research was carried out we realized that the same approach was also used in the literature [7, 8]. While we in this paper use the mutual information as a measure of goodness of an attribute, the previous techniques used a t-test [7] and the information gain ratio [8] as the measure.

Our main results and contributions can be listed below:

- We propose Rank Forest to primarily overcome the potential shortcomings generated by the randomness property of the Random Forest.
- Our proposed algorithm not only shows better ensemble accuracy than Random Forest but also builds tree predictors that are more concrete, stable, robust towards errors, as well as less dependent on the number of tree predictors.
- We conduct experiments to show that the proposed method performs well in term of ensemble accuracy, average individual tree accuracy of a decision forest, and average time to construct a tree predictor. Moreover, the new method is computationally less expensive for tree construction in ensemble learning.

The rest of the paper is organized as follows. Section 2 covers literature review regarding well-known decision forest algorithms, decision tree algorithms and feature ranking techniques. Section 3 discusses our approach called Rank Forest. Section 4 consists of our results and comparisons with Random Forest algorithm. Section 5 concludes our paper.

2 Literature Review

Current section has been divided into three different parts: part 1 discusses the literature regarding different decision tree algorithms, part 2 discusses different decision forest algorithms. While part 3 states different feature/attribute ranking techniques (we will use to assign weights to each attribute during feature bagging).

2.1 Decision Tree Algorithms

There are numerous decision tree algorithms have been proposed in literature [9–11]. However, in this section we will discuss latest algorithms to build decision tree.

ID3: It is one of the earliest decision tree algorithms proposed by Quinlan in [12]. This technique builds decision tree on a dataset containing only nominal or categorical attributes. In ID3 algorithm, in order to make a split at each node it calculates Information Gain based gain ratio for each of the attributes. The one with highest gain ratio will be chosen to make a split at this node. Data will be divided on the base of unique values of the decision node. Same procedure will be applied on each subset of the data after division. The process is purely recursive until it finds leaf.

C4.5: It is one of the latest algorithms to build a decision tree, presented by Quinlan in [11]. It is an extension of Quinlan's previously presented technique called ID3. In this algorithm, in order to determine a splitting attribute, it uses information theoretic based normalized information gain to find the appropriate attribute as a decision attribute. Significant improvements had been made in C 4.5 as compare to ID3, (1) ID3 can only handle nominal attributes while C 4.5 has capacity to handle attributes with continuous values as well (2) It possess capacity to handle noise and missing values in data (3) It does pruning in built tree to overcome the overfitting issues in the model. (4) Moreover, it can generate rules on the base of generated tree in order to classify unknown samples.

Explore: This technique has been presented by Islam et al. in [10]. It is one of the extension of C 4.5. In this technique, a new attribute ranking criteria Ultimate Gain Ratio has been used instead of Gain Ratio (used in C 4.5). This technique introduces new criteria to select splitting value of decision attribute (both in case of numerical or categorical). In case of numerical attribute, it uses best window of values as splitting point instead of choosing a single value. In case of categorical attribute, it makes a single attribute value as splitting value, unlike C 4.5 where we make branches against each of the unique value of the decision categorical attribute.

2.2 Decision Forest Algorithms

Several decision forests algorithms have been presented in literature, that demonstrate variety regarding the use of training data or feature set. In this section, we discuss some of the major decision forest algorithms along with their limitations.

Bagging [13] is one of the pioneer decision forest algorithms. This algorithm does not utilize whole training data but generates a new training data set D_x from the original training data set D using bootstrap sampling [14]. D_x contains same number of records as D but these records are chosen randomly with replacement from D. In Bagging, we generate K bootstrap samples and learn a tree model using each of the bootstrap samples. Total number of K *or* |T| tree models are built to predict test data set. Voting strategy or Out-Of-Bag error is used for performance analysis of the model. For data sets containing similar (i.e. difference is low) records, bagging may not extract strong diverse records, since it extracts sample records with replacement. Moreover, Bagging is sort of computationally expensive as compare to the other decision forest algorithms since it uses full feature or attribute space for each tree predictor.

Random Subspace [15] is another decision forest algorithm which differs from bagging regarding the use of training data. It utilizes whole training data for each tree of the forest. Unlike bagging, in order to determine the splitting attribute for each node, Random Subspace randomly draws a subset s (subspace) of attributes or a sample of attributes from the entire attribute set A.

$$|s| = \text{int}(\log_2 |A|) + 1 \tag{1}$$

Random Forest is one of the major decision forest algorithms which is a conglomerate of both Bagging and Random Subspace. In fact, it considered as one of the most acknowledged state of the art decision forest algorithms being in use. Instead of applying Random Subspace on the original training data set, Random Forest inputs a different bootstrap sample D_x for each tree learner.

Size of s is calculated on the base of criteria given in Eq. (1) [5]. The size of s as a ratio of n determines the diversity among different tree learners, for Instance if |A| or n is 3 then calculated |s| is 3 which is almost 100% of the total $n = 3$. In this case, |s| is near about n; hence we cannot expect much more diversity among different tree learners of the forest. On the other hand, if |A| is large say, 150 then |s| contains 8 randomly chosen attributes ($\text{int}(\log_2 150) + 1 = 8$) covering only 5% of the total attributes. A lower number of attributes also reduces the chance of containing adequate number of attributes with high discriminative capability among different classes. Lower individual accuracy of the trees can be observed if the number of discriminative attributes is not high enough in s. In order to handle this issue, in this paper we use feature ranking techniques to assign some weights to each feature as per its discriminative capability during feature bagging. This paper proposes our method Rank Forest in order to handle the above stated issue.

Adnan et al. [16] present a new decision forest technique called Forest CERN. This technique aims to improve potential issues that can be observed in CS4 [17] and SysFor [18]. This technique utilizes potential classification capacity of all non-class attributes, unlike CS4 and SysFor where a subset of non-class attributes has been used as root nodes of tree predictors. Core idea behind this technique is to create diversity among tree predictors by excluding the attributes that acted as root node in previous trees. This technique imposes some penalty on recently used root attribute in order to hinder it for reappearing as root node in subsequent trees. As the algorithm makes progress it gradually revokes penalty to increases the probability of reappearing in subsequent trees.

2.3 Feature Ranking Techniques

Numerous techniques have been used in order to rank features with respect to their capacity to discriminate among different categories. Here we discuss two of the most popular base line feature ranking techniques.

Mutual Information. Mutual information is one of the criteria use to quantify the information that can be obtained from an attribute A_i in the absence of attribute C [19]. In other words, it describes how much information an attribute can provide to predict a class. Mutual Information score of an attribute A_i with respect to C can be calculated as follow

$$I(A_i, C) = \sum_{c \in C} \sum_{x \in A_i} Pr(c, x) \log\left(\frac{Pr(c, x)}{Pr(c)p(x)} \right) \tag{2}$$

Information Gain. Information Gain is also one of the attribute/feature goodness criteria, which measures the amount of information can be obtained for category prediction by knowing the presence or absence of an attribute value in a data instance. It mainly uses entropy to validate the discriminative capability of an attribute. Information Gain (IG) of an attribute with respect to class attribute C can be calculated as follow.

$$G(A_i) = -\sum_{i=1}^{n} P_r(c_i) \log P_r(c_i) + P_r(A_i) \sum_{i=1}^{n} P_r(c_i|A_i) \log P_r(c_i|A_i)$$
$$+ P_r(\bar{A}_i) \sum_{i=1}^{n} P_r(c_i|\bar{A}_i) \log P_r(c_i|\bar{A}_i) \tag{3}$$

3 Rank Forest

In this section, our proposed algorithm Rank Forest has been described and elaborated in detail.

Rank Forest has been built on the top of Random Forest. It is also a mixture of Bagging and Random Subspace. However, it does not utilize same strategy for subspace selection. Since in Random Forest if the ratio of $|s|$ with n is much less than n then the chances of the selection of attributes with higher discrimination capability get reduced. This downside of the Random Forest potentially affects the overall accuracy of the model. In Rank Forest, we propose to handle it by distributing non- uniform probability weights (calculated using Mutual Information score of that attribute) over attribute set. These weights will act as non- uniform probabilities for random sampling of attributes subspace. Pseudo code for Rank Forest is given below in Algorithm 1.

Algorithm 1: *Rank Forest*

Input: Data matrix D as Training dataset, D_T as Test dataset, T as set of tree predictors and K is the number of tree predictors.

Output: Ensemble of Trees

1) Calculate Mutual Information MI score $I(A_i; C)$ against each of the feature/attribute vector $A_i \in D$ and store the MI scores in W, where $W = \{w_1, w_2, w_3 ..., w_n\}$, w_i represents Non-normalized MI score/weight for attribute A_i

2) Range normalization of weights between $(0, 1)$ by dividing each weight with the sum of W and store it in W_{norm}.

$$W_{norm} = \frac{w_{\{i=1...n\}}}{sum(W)}$$

3) Initialize sample size $N = m$, here

For $i = 1$ to K

    i.   Draw a bootstrap sample $Train_s$ of size N from Training data D

    ii. Grow a forest tree T_i over the bootstrapped data, by recursively repeating following steps for each terminal node of the tree until min size reached

        b. Select $|s|$ variables at random from n original variables, with attached selection probability distribution as W_{norm}

        c. Pick the best variable/ split- point among s

        d. Split the node into two daughter nodes

4) Output ensemble of trees as $\{T_i\}_1^K$

5) To make the class prediction for $x \in Test$

    i.   Let suppose $TreePredict_p(x)$ prediction of class label by pth tree for the data instance x of the Test dataset

    ii. $ForestPredict(x) = majorityVote\{TreePredict_p(x)\}^K_{p=1}$, prediction by forest for data instance x of the Test dataset

End

Step 1, calculates the discriminative scores of each attribute using Mutual Information criteria and considers these scores as non- normalized weights W.

Step 2, performs range normalization of given weights W to convert score values into probabilistic values (sum of all values must be 1) and stores normalized weights as W_{norm}. Attribute with higher MI score will get higher probability weight.

Step 3, applies Rank Forest algorithm (mutated Random Forest) on Training data, taking each bootstrap sample of size N. To make a split at each node it takes random sample of attributes with non-uniform probability distribution as W_{norm}, unlike Random Forest where it draws sample with uniform probabilities. Non-uniform probability distribution will increase the chances that attributes with higher discriminative power get selected (discussed in Sect. 2.1). The rest of procedure is same as in Random Forest.

4   Experiments and Results Analysis

In order to evaluate the proposed method, as compared to well- known Random Forest algorithm, we come up with a unique and an empirical analysis of these algorithms. We evaluate our proposed method on nine different datasets (from UCI machine learning repository and KEEL datasets repository) using three evaluation measures (ensemble Accuracy, average individual accuracy for tree predictors in decision forest, and average time to construct single tree predictor). Accuracies of each model, including proposed model, have been calculated using confusion matrix. Computational time has also been recorded to assess the computational efficiency of the model.

4.1 System Specification

All of the experiments reported in this paper are calculated using machine with Intel (R) Core i7, 2.4 GHz of Processor, with 4 GB RAM running under 64 bit Microsoft Windows 10 operating system. Primarily, Python programming language has been used for implementation purpose.

4.2 Evaluation Measures

To evaluate our proposed approach in comparison to other models, the *accuracy* score has been used as a performance measure. It is basically the degree of agreement of the measurement of a quantity to actual value of that quantity. It is defined as the percentage of number of correctly classified data to total classified data. Given confusion matrix for a 2-class problem in Table 1.

Table 1. Confusion matrix

	Predictive positive	Predictive negative
Actual positive	a	b
Actual negative	c	d

where a is the number of positive instances correctly classified as positive, b is the number of positive instances incorrectly classified as negative, c is the number of negative instances classified as positive, and d is the number of negative instances correctly classified as negative.

We can compute the accuracy measure between predicted and actual measure as

$$Accuracy = \frac{a + d}{a + b + c + d} \qquad (4)$$

In this paper, we evaluate each algorithm on three different measures Ensemble Accuracy (using accuracy measure), time to construct a tree with respect to given dataset, and average individual tree accuracy in decision forest.

4.3 Benchmark Datasets

In our experimental work, we use nine datasets. These datasets have been taken from UCI Machine Learning repository[1] and KEEL Datasets repository[2]. These datasets have been considered to assess performance over a variation in number of features (for instance 8, 13, ..., 36, 60, and 90) and number of classes (for instance 2, 3, 8, and 15). The detail for each dataset has been given in Table 2.

Table 2. Detail of each dataset

Dataset	# of attributes	# of instances	# of classes
Movement Libras	90	360	15
Sonar	60	208	2
Splice	60	5620	3
Chess	36	3196	2
Ionosphere	33	351	2
WDBC	30	569	2
German credit	20	1000	2
Wine	13	178	3
Ecoli	8	336	8

4.4 Experimental Setup

Experiments have been reported with stratified 10- fold cross validation. In 10-fold cross validation, we divide dataset into ten equal segments. Each time we take one segment as a test dataset and remaining nine segments as training dataset. Hence, a total of ten training and ten test datasets has been generated in general. In case of decision forest algorithm, we use K tree predictor under the hood of ensemble learning. In case of 10-fold cross validation, for each fold we generate $K = 100$ tree and for a total of 10 folds we generate 1000 trees. For each fold, we evaluate the performance of tree predictor by using respective test segment. Final accuracy will be calculated as an average of accuracies of test datasets in all 10 folds.

4.5 Results

In this section, we evaluate our proposed method comparatively, in terms of accuracy and running time. Primarily, we compare our results with Random Forest because our prime aim is to overcome the downsides of Random Forest algorithm. As it has been discussed in Sect. 2.1, Random forest is one of the techniques aims to handle instability issue as well as computational cost issue in decision tree, under the hood of ensemble learning. However, the randomness property of Random Forest might overshadow the consistency of results. Our proposed technique Rank Forest, being an extension to Random Forest, aims to handle downsides of randomness property of Random Forest.

---

[1] http://archive.ics.uci.edu/ml/.
[2] http://sci2s.ugr.es/keel/category.php?cat=clas.

Ensemble Accuracy is one of the key measures to assess the performance of decision forest algorithms. Decision forest approach has been adopted to overcome the instability issues in decision tree. It takes an ensemble of results from multiple tree predictors. In Table 3, we present a comparison of ensemble accuracies of Rank Forest and Random Forest. In order to overcome the biased behavior, all of the reported results are an average of 10-fold cross validation. Standard deviation among 10 segments of cross validation with respect to each algorithm has also been reported. On the base of accuracy with respect to particular dataset, we assign ranks (1 to 2, 1 as best and 2 as worst) to each of the models in comparison.

Table 3. Ensemble accuracies

Dataset	Rank Forest	Random Forest
Chess	**0.95 ± 0.03 (1)**	**0.95 ± 0.02 (1)**
WDBC	**0.97 ± 0.02 (1)**	0.95 ± 0.03 (2)
German credit	**0.77 ± 0.04 (1)**	0.76 ± 0.03 (2)
Splice	0.96 ± 0.01 (2)	**0.97 ± 0.005 (1)**
Ionosphere	**0.89 ± 0.07 (1)**	0.88 ± 0.08 (2)
Movement Libras	0.78 ± 0.15 (2)	**0.79 ± 0.14 (1)**
Wine	**0.51 ± 0.10 (1)**	0.49 ± 0.07 (2)
Ecoli	**0.85 ± 0.04 (1)**	**0.85 ± 0.04 (1)**
Sonar	**0.73 ± 0.1 (1)**	0.72 ± 0.1 (2)
Average accuracy	**0.82 (1)**	**0.81(2)**
Best performer in number of datasets	**7 (Rank 1st)**	**4 (Rank 2nd)**

In Table 3, we can observe that Rank Forest outperforms Random Forest in majority of the datasets. Here we can see that out of 9 datasets Rank Forest dominates the Random Forest in 5 datasets with average ensemble accuracy as 0.82. In two datasets both the algorithms perform in a similar accuracy so Rank Forest's best performance can be seen in 7 (5 + 2) datasets. Since in case of chess and ecoli datasets both the algorithms perform similar, therefore we do not consider it as dominance. The dominance rate of Rank Forest is (5/7) 0.72 therefore it secures 1st rank as compare to Random Forest algorithm. On the other hand, Random Forest leads the performance in 2 out of 9 datasets (For Instance, Movement Libras and Splice). Since in two datasets it equates the performance with Rank Forest so in actual it remained best performer in 4 (2 + 2) datasets. But we do not consider equal performance as dominance. Although Random Forest is much closer to the Rank Forest in term of average ensemble accuracy (0.81) yet its dominance rate across the datasets is 0.28 (2/7) (excluding the chess and ecoli datasets) which is much lower than Rank Forest. Furthermore, Rank Forest and Random Forest resemble in performance with minute difference but still Rank Forest stands healthier and more improved than Random Forest in term of ensemble accuracy.

Higher ensemble accuracy can be obtained only if we have balance individual tree accuracy in ensemble learning. A tree predictor with better accuracy also indicates that it considered attributes with better discriminative capability.

Moreover, a decision forest with better individual tree accuracy will have better tree structure and ultimately more robust towards error. In Table 4, we present average of reported accuracies for individual trees in ensemble learning algorithms with respect to each dataset.

Table 4. Average individual tree accuracies

Dataset	Rank Forest	Random Forest
Chess	0.92 ± 0.04 (1)	0.85 ± 0.03 (2)
WDBC	0.93 ± 0.01 (1)	0.93 ± 0.01 (2)
German credit	0.68 ± 0.01 (1)	0.67 ± 0.01 (2)
Splice	0.92 ± 0.009 (1)	0.82 ± 0.009 (2)
Ionosphere	0.85 ± 0.07 (1)	0.85 ± 0.07 (2)
Movement Libras	0.47 ± 0.1 (1)	0.46 ± 0.1 (2)
Wine	0.46 ± 0.03 (1)	0.45 ± 0.02 (2)
Ecoli	0.77 ± 0.02 (1)	0.71 ± 0.02 (2)
Sonar	0.64 ± 0.06 (1)	0.62 ± 0.04 (2)
Average accuracy	**0.74 (1)**	**0.70 (2nd)**
Best performer in number of datasets	**9 (Rank 1st)**	**0 (Rank 2nd)**

In above reported results, we observe that Rank Forest secures 1st rank across the line of all datasets with average individual tree accuracy 0.74. Rank Forest outperforms Random Forest in all the 9 datasets. It reports dominance rate in given datasets as 100%, since it remains best in 9 out of 9 datasets. Random forest, in terms of lead in number of datasets, performs worst, since it does not get lead in any of the given datasets. Overall it gets individual tree accuracy as 0.70 which is marginally lower than Rank Forest's reported accuracy. Hence Rank Forest again outperforms the Random Forest and selects attributes with better discriminative capability.

Observing the performance of both models in term of individual tree accuracy we see that Rank Forest builds tree predictors that are more concrete and robust towards error than tree build by Random Forest. Poor performance of Random Forest in term of individual tree accuracy clearly indicates that Random Forest's performance is heavily dependent on voting strategy or collective performance of all the tree predictors rather than the performance of individual tree predictor. Moreover, we can also conclude that Random Forest performance (ensemble accuracy) is purely directly proportional to the value of K (number of tree predictors in voting), greater the number of tree predictors higher will be the ensemble accuracy for Random Forest. For instance, to check this behavior we notice that If we run Random Forest algorithm on a range of values of K (For instance, 100, 75, 50, 30, and 10) using the *Splice* dataset (Random Forest performed better over Rank Forest in term ensemble accuracy using this dataset) we can see a decline of performance by Random Forest from 0.97 to 0.95 (range 100 to 30). On the other hand, Rank Forest reports stable performance (ensemble accuracy 0.96) despite the decrease in number of tree predictors (from 100 to 30). When K gets value under 30 then the ensemble accuracy for Rank Forest also starts bending. Hence, we can conclude that Rank Forest not only performs better in term of ensemble accuracy but also builds tree predictors those are more concrete and robust towards error (less dependency on the value of K).

Third and the most important evaluation measure that we consider is the amount of time (average time in seconds) a decision forest model takes to build a tree, with given bootstrap sample and attribute set. Here, we show average time tree predictors (1000 trees, 10-fold and in each fold 100 tree predictor) consume to construct a single tree over particular sample dataset. In Table 5, we present average time taken (in second) by each decision forest model to construct a tree predictor along with given rank (1 to 2, lower the time higher the rank).

Table 5. Average time (in second) to build a tree predictor

Dataset	Rank Forest	Random Forest
Chess	0.2 ± 0.01 (1)	0.3 ± 0.01 (2)
WDBC	1.42 ± 0.06 (2)	1.14 ± 0.04 (1)
German credit	0.6 ± 0.01 (2)	0.3 ± 0.01 (1)
Splice	0.62 ± 0.01 (1)	0.9 ± 0.02 (2)
Ionosphere	2.87 ± 0.2 (2)	2.81 ± 0.3 (1)
Movement Libras	1.4 ± 0.07 (2)	1.36 ± 0.06 (1)
Wine	0.48 ± 0.02 (2)	0.43 ± 0.02 (1)
Ecoli	0.17 ± 0.01 (2)	0.12 ± 0.004 (1)
Sonar	0.35 ± 0.008 (1)	0.35 ± 0.01(1)
Average time	0.9	0.86
Best performer in number of datasets	3	7

In below reported times, it can be observed that Random Forest as a best gets 1st rank in 6 out of 9 datasets with average time 0.86 s. Rank Forest secures 1st rank in 2 datasets, moreover it shares 1st rank with Random Forest in one other dataset. Overall Rank Forest as an average consumes 0.9 s to construct a single tree predictor. We may notice a minute difference in time consumption (0.86 and 0.9 s) by Random Forest and Rank Forest. Rank Forest and Random Forest resemble in construction procedure but Rank Forest gets overall 2nd best performer in term of time. The only reason we can find is, during random feature selection, Random Forest considers uniform probability distribution while rank forest considers non- uniform distribution. Hence random sampling with uniform distribution takes less time than non- uniform distribution. Therefore, Random Forest is slightly more efficient than Rank Forest in term of time.

Now let's evaluate each model on the base of steady performance in all the three evaluation measures stated above (Ensemble Accuracy, Average Individual Tree accuracy, and Time to construct a single tree). In Table 6, we report on the evaluation scores of both decision forest models (including proposed model) with respect to three evaluation measures. The purpose behind putting all the scores in one table is to assess the balance behavior of each model. Let's have a look at performance measures of each model one by one in Table 6.

Table 6. Reported values for three evaluation measures

	Ensemble accuracy rank	Individual tree accuracy rank	Average time to construct a tree predictor
Rank Forest	1st	1st	**0.9 (s)**
Random Forest	2nd	2nd	0.86 (s)

In case of Random Forest, we can observe that it gets 2nd best rank in both the accuracy measures (ensemble accuracy and individual tree accuracy) following Rank Forest. However, in term of time it consumes a comparable but less time than best performer Rank Forest algorithm. Here, we see that time consumption difference between both models is very minute. Since in previous section we discussed that Rank Forest has capability to achieve better performance with low population of tree predictors which is not the case with Random Forest algorithm. Therefore, the difference of time can be overcome with lower population of tree predictors. So here it is evident that Rank Forest not only stands as best and stable performer in both accuracy measures but also consumes time which is marginal to Random Forest.

5   Conclusion

In this paper, we proposed a new decision forest building method, called Rank Forest, to overcome some major issues of decision trees, namely instability and computational cost. Random Forest first randomly selects a subset of attributes from a full set of attributes and then selects the best attribute from the sample subset for a decision node. In Rank Forest, we draw a random subset of attributes from a full set of attributes with non- uniform probability distribution. The weight attached to each attribute depicts its capability to discriminate among different classes. The non- uniform weights of attributes increase the chance of the selection of a good attribute. An attribute with a higher discriminative score will get a higher probability value than an attribute with a lower discriminative score. Hence, a good attribute will have a high chance of getting selected in the course of random selection. The same approach was also taken by some existing techniques [7, 8], but the measures for computing the goodness of attributes are different in each of these techniques. It can be an interesting future work to explore whether or not there is any difference in ensemble accuracy and individual tree accuracy due to the selection of different measures. In this paper, we evaluated our proposed method in three different ways (ensemble accuracy, individual tree accuracy, and average time to construct tree predictor) on nine different datasets. Results of our experiments show that Rank Forest typically achieves higher ensemble accuracy and individual accuracy than Random Forest. In term of running time, Rank Forest takes comparable construction time as Random Forest.

References

1. Feldman, R., Sanger, J.: The Text Mining Handbook: Advanced Approaches in Analyzing Unstructured Data. Cambridge University Press, Cambridge (2006)
2. Martin, J.K., Hirschberg, D.S.: The time complexity of decision tree induction (1995)
3. Chikalov, I.: Average Time Complexity of Decision Trees. Intelligent Systems Reference Library, vol. 12. Springer, Heidelberg (2011). https://doi.org/10.1007/978-3-642-22661-8
4. Dietterich, T.G.: Ensemble methods in machine learning. In: Kittler, J., Roli, F. (eds.) MCS 2000. LNCS, vol. 1857, pp. 1–15. Springer, Heidelberg (2000). https://doi.org/ 10.1007/3-540-45014-9_1
5. Breiman, L.: Random forests. Mach. Learn. **45**, 5–32 (2001). https://doi.org/10.1023/A: 1010933404324
6. Biau, G., Biau, G.: Analysis of a random forests model. J. Mach. Learn. Res. **13**, 1063–1095 (2012)
7. Amaratunga, D., Cabrera, J., Lee, Y.-S.: Enriched random forests. Bioinformatics **24**, 2010–2014 (2008). https://doi.org/10.1093/bioinformatics/btn356
8. Zhao, H., Williams, G.J., Huang, J.Z.: wsrf: an R package for classification with scalable weighted subspace random forests (2017). jstatsoft.org
9. Hssina, B., Merbouha, A., Ezzikouri, H., Erritali, M.: A comparative study of decision tree ID3 and C4.5. Int. J. Adv. Comput. Sci. Appl. (IJACSA) **4**(2) (2014)
10. Islam, M.Z.: EXPLORE: a novel decision tree classification algorithm. In: MacKinnon, L.M. (ed.) BNCOD 2010. LNCS, vol. 6121, pp. 55–71. Springer, Heidelberg (2012). https:// doi.org/10.1007/978-3-642-25704-9_7
11. Quinlan, J.R.: C4.5: Programs for Machine Learning. Morgan Kaufmann Publishers, Los Altos (1993)
12. Quinlan, J.R.: Induction of decision trees. Mach. Learn. **1**, 81–106 (1986)
13. Breiman, L.: Bagging predictors. Mach. Learn. **24**, 123–140 (1996)
14. Han, J., Kamber, M.: Data Mining: Concepts and Techniques. Morgan Kaufmann Publishers, Los Altos (2006)
15. Ho, T.K.: The random subspace method for constructing decision forests. IEEE Trans. Pattern Anal. Mach. Intell. **20**, 832–844 (1998). https://doi.org/10.1109/34.709601
16. Adnan, M.N., Islam, M.Z.: Forest CERN: a new decision forest building technique. In: Bailey, J., Khan, L., Washio, T., Dobbie, G., Huang, J.Z., Wang, R. (eds.) PAKDD 2016. LNCS (LNAI), vol. 9651, pp. 304–315. Springer, Cham (2016). https://doi.org/ 10.1007/978-3-319-31753-3_25
17. Li, J., Liu, H.: Ensembles of cascading trees. In: Proceedings of the Third IEEE International Conference on Data Mining, pp. 585–588 (2003)
18. Islam, M.Z., Giggins, H.: Knowledge discovery through SysFor -a systematically developed forest of multiple decision trees. In: Proceedings of the 9th Australasian Data Mining Conference (2011)
19. Xu, Y., Jones, G., Li, J., Wang, B., Sun, C.: A study on mutual information-based feature selection for text categorization. J. Comput. Inf. Syst. **3**(3), 203–213 (2005)

Performance Evaluation of a Distributed Clustering Approach for Spatial Datasets

Malika Bendechache[1(✉)], Nhien-An Le-Khac[2], and M-Tahar Kechadi[1]

[1] Insight Centre for Data Analytics, University College Dublin,
O'Brien Building, Centre East, Belfield, Dublin 04, Ireland
`malika.bendechache@gamil.com, tahar.kechadi@ucd.ie`
[2] University College Dublin, Belfield, Dublin 04, Ireland
`an.lekhac@ucd.ie`

Abstract. The analysis of big data requires powerful, scalable, and accurate data analytics techniques that the traditional data mining and machine learning do not have as a whole. Therefore, new data analytics frameworks are needed to deal with the big data challenges such as volumes, velocity, veracity, variety of the data. Distributed data mining constitutes a promising approach for big data sets, as they are usually produced in distributed locations, and processing them on their local sites will reduce significantly the response times, communications, etc. In this paper, we propose to study the performance of a distributed clustering, called Dynamic Distributed Clustering (DDC). DDC has the ability to remotely generate clusters and then aggregate them using an efficient aggregation algorithm. The technique is developed for spatial datasets. We evaluated the DDC using two types of communications (synchronous and asynchronous), and tested using various load distributions. The experimental results show that the approach has superlinear speed-up, scales up very well, and can take advantage of the recent programming models, such as MapReduce model, as its results are not affected by the types of communications.

Keywords: Distributed data mining · Distributed computing
Synchronous communication · Asynchronous communication
Spacial data mining · Super-speedup

1 Introduction

Nowadays big data is becoming a commonplace. It is generated by multiple sources at rapid pace, which leads to very large data volumes that need to be stored, managed, and analysed for useful insights. From organisations point of view, it is not the size of the generated data which is important. It is what we learn from it that matters, as this may help understanding the behaviour of the system that is governed by this data or help to make some key decisions, etc. To extract meaningful value from big data, we need appropriate and efficient mining

© Springer Nature Singapore Pte Ltd. 2018
Y. L. Boo et al. (Eds.): AusDM 2017, CCIS 845, pp. 38–56, 2018.
https://doi.org/10.1007/978-981-13-0292-3_3

and analytics techniques to analyse it. One of the most powerful and common approaches of analysing datasets for extracting useful knowledge is clustering. Clustering has a wide range of applications and its concept is so interesting that numerous algorithms for various types of data have been proposed and implemented. However, big data come up with new challenges, such as large volumes, velocity, variety, and veracity, that the majority of popular clustering algorithms are inefficient at very large scale. This inefficiency can be that the final results are not satisfactory or the algorithm has high complexity which requires large computing power and response time to produce final results. There are two major categories of approaches to deal with computational complexity of these clustering algorithms: (1) the first category consists of reducing the size of the initial dataset. One can use either sample-based techniques or dimensionality reduction techniques. The second category consists of using parallel and distributed computing to speed up the response time. In this case we can try to parallelise or model the algorithm in the form of a client-server model using MapReduce mechanism. However, these algorithms are inherently difficult to parallelise, and designing an efficient distributed version of the algorithm is not straightforward either. This is due to the fact that the processing nodes, either in the parallel or distributed versions, need to communicate and coordinate their efforts in order to obtain the same results. These communications are extremely expensive and can cancel the benefit of the parallelised version. To deal with these challenging issues, we propose to study a distributed approach that takes advantage of parallel and distributed computing power, while getting ride of the drawbacks of the previous methods. In addition, one of the main advantages of our approach is that it can be used as a framework for all clustering algorithms. In other words, while it is well known that there is no clustering algorithm that can universally be used to cluster every dataset of a given application, our approach can be used for all algorithms or a set of algorithms to derive a distributed clustering approach for a given data having specific characteristics.

The proposed approach has two main phases: the first phase, based on the SPMD (Single Program Multiple Data) paradigm, consists of dividing the datasets into K partitions, where K is the number of processing nodes. Then, for each partition we cluster its data into C_i clusters. This phase is purely parallel, as each processing node executes a clustering algorithm on its data partition independently of the others. The obtained clusters on each node are called local clusters. This phase does not require any communications, and in addition, in the majority of applications the data is collected by various sources, which are geographically distributed. Therefore, the data is already partitioned. All required is to cluster locally the data. The second phase consists of aggregating (or merging) the local clusters to obtain global clusters by merging overlapping clusters. In order to determine whether two local clusters belonging to two different nodes are overlapping or not, one needs to exchange the local clusters between the nodes. This operation is extremely expensive when the dataset is very large. The main idea of our approach is to minimise the data exchange while maximising the quality of the global clusters. The method used to aggregate spatial local clusters into global clusters allows only to exchange about 2% of the original datasets [28], which is highly efficient. In this paper, we want to study the performance of such distributed clustering technique by calculating its speedup compared to the sequential version of the algorithm, its scalability, its communication overheads, and its complexity in general.

The rest of the paper is organised as follows: In the next section we will give an overview of the state-of-the-art of parallel and distributed data mining techniques and discuss their limitations. Then we will present in more details the proposed distributed framework and its concepts in Sect. 3. In Sect. 4, we evaluate its performance based on two types of implementations; synchronous and asynchronous communications. In Sect. 5, we discuss the experimental results based on speedup, scalability, communication overheads, and compare the two implementation models; synchronous and asynchronous. Finally, we conclude in Sect. 6.

2   Related Work

Distributed Data Mining (DDM) is a line of research that has attracted much interest in recent years [25]. DDM was developed because of the need to process data that can be very large or geographically distributed across multiple sites. This has two advantages: first, a distributed system has enough processing power to analyse the data within a reasonable time frame. Second, it would be very advantageous to process data on their respective sites to avoid the transfer of large volumes of data between the site to avoid heavy communications, network bottlenecks, etc.

DDM techniques can be divided into two categories based on the targeted architectures of computing platforms [36]. The first, based on parallelism, uses traditional dedicated and parallel machines with tools for communications between processors. These machines are generally called super-computers. The second category targets a network of autonomous machines. These are called distributed systems, and are characterised by distributed resources, low-speed network connecting the system nodes, and autonomous processing nodes which can be of different architectures, but they are very abundant [23]. The main goal of this category of techniques is to distribute the work among the system nodes and try to minimise the response time of the whole application. Some of these techniques have already been developed and implemented in [1,34].

However, the traditional DDM methods are not always effective, as they suffer from the problem of scaling. One solution to deal with large scale data is to use parallelism, but this is very expensive in terms of communications and processing power. Another solution is to reduce the size of training sets (sampling). Each system node generates a separate sample. These samples will be analysed using a single global algorithm [26,37]. However, this technique has a disadvantage that the sampling in this case is very complex and requires many communications between the nodes which may impact on the quality of the samples and therefore the final results. This has led to the development of techniques that rely on ensemble learning [4, 32]. These new techniques are very promising, as each technique of the ensemble network attempts to learn from the data and the best or compromised results of the network will emerge as the winner. Integrating ensemble learning methods in DDM framework will allow to deal with the scalability problem, as it is the case of the proposed approach.

Clustering algorithms can be divided into two main categories, namely partitioning and hierarchical. Different elaborated taxonomies of existing clustering algorithms are given in the literature. Many parallel clustering versions based on these algorithms have been proposed in the literature [2, 14, 14, 17, 21, 22, 35]. These algorithms are further classified into two sub-categories. The first consists of methods requiring multiple rounds of message passing. They require a significant amount of synchronisations and data exchange. The second sub-category consists of methods that build local clustering models and send them to a central site to build global models [28].

In [14, 15], message-passing versions of the widely used K-Means algorithm were proposed. In [17, 35], the authors dealt with the parallelisation of DBSCAN; density-based clustering algorithm. In [21] a parallel message passing version of the BIRCH algorithm was presented. A parallel version of a hierarchical clustering algorithm, called MPC for Message Passing Clustering, which is especially dedicated to Microarray data was introduced in [22]. Most of the parallel approaches need either multiple synchronisation constraints between processes or a global view of the dataset, or both [2]. All these approaches deal with the parallelisation of the sequential version of the algorithm by trying phases of the algorithm which can be executed in parallel by several processors. However, this requires many synchronisations either to access shared data (for the shared memory model) or communications (for message passing model). In some algorithm these synchronisations and communications are extremely expensive and it is not worth parallelising them. This approach is not usually scalable.

In [9] a client-server model is adopted, where the data is equally partitioned and distributed among the servers, each of which computes the clusters locally and sends back the results to the master. The master merges the partially clustered results to obtain the final results. This strategy incurs a high communication overhead between the master and slaves, and a low parallel efficiency during the merging process. Other parallelisations using a similar client-server model include [3, 11, 12, 20, 24, 38]. Among these approaches, various programming mechanisms have been used, for example, a special parallel programming environment, called skeleton based programming in [12] and parallel virtual machine in [24]. A Hadoop-based approach is presented in [20].

Another approach presented in [2] also applied a merging of local models to create the global models. Current approaches only focus on either merging local models or mining a set of local models to build global ones. If the local models cannot effectively represent local datasets then global models accuracy will be very poor [28]. In addition, both partitioning and hierarchical categories have some issues which are very difficult to deal with in parallel versions. For the partitioning class, it needs the number of clusters to be fixed in advance, while in the majority of applications the number of classes is not known in advance. For the hierarchical clustering algorithms, they have the issue of stopping conditions for clustering decomposition, which is not an easy task and mainly in distributed versions.

3   Dynamic Distributed Clustering

Dynamic Distributed Clustering (DDC) model is introduced to deal with the limitations of the parallel and master-slave models. DDC combines the characteristics of both partitioning and hierarchical clustering methods. In addition, it does neither inherit the problem of the number of partitions to be fixed in advance nor the problem of stopping conditions. It is calculated dynamically and generates global clusters in a hierarchical way. All these features look very promising and some of them have been thoroughly studies in [7], such as the dynamic calculation of the number of the clusters and the accuracy of the final clustering, in this study one wants to show the effect of the communications on the response time, the communication model used, the scalability of the approach, and finally its performance in terms of speed up compared to the sequential version. In this paper we will focus on

- Synchronous and asynchronous communications, as this approach can be implemented either with synchronous or asynchronous communications. Both implementations produce the same results.
- The speed-up of the DDC approach using DBSCAN as the basic algorithm for clustering the partitions. This algorithm is known to have non-polynomial complexity $(O(n^2))$.
- Scalability of the approach as the size of the dataset increases.

We start by briefly explaining the algorithm and then present a performance and evaluation model for the approach.

The DDC approach has two main phases. In the first phase, we cluster the datasets located on each processing node and select good local representatives. All local clustering algorithms are executed in parallel without communications between the nodes. As DBSCAN is the basic algorithm for clustering local datasets, we can reach a super linear speed-up of p^2, where p is the number of processing nodes. The second phase collects the local clusters from each node and affects them to some special nodes in the system; called leaders. The leaders are elected according to their characteristics such as capacity, processing power, connectivity, etc. The leaders are responsible for merging the local clusters. In the following we explain how the local clusters are represented and merged to generate global clusters.

3.1   Local Models

The local clusters are highly dependent on the clustering techniques used locally in each node. For instance, for spatial datasets, the shape of a cluster is usually

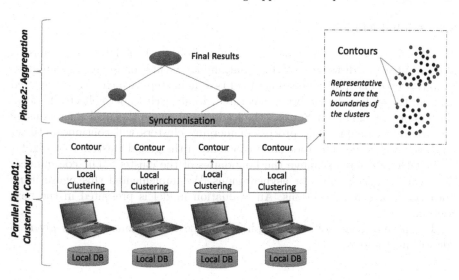

Fig. 1. An overview of the DDC approach.

dictated by the technique used to obtain them. Moreover, this is not an issue for the first phase, as the accuracy of a cluster affects only the local results of a given node. However, the second phase requires sending and receiving all local clusters to the leaders. As the whole data is very large, this operation will saturate very quickly the network. So, we must avoid sending all the original data through the network. The key idea of the DDC approach is to send only the cluster's representatives, which constitute between 1% and 2% of the whole data. The cluster representatives consist of the internal data representatives plus the boundary points of the cluster.

There are many existing data reduction techniques in the literature. Many of them are focusing only on the dataset size. For instance, they try to reduce the storage capacity without paying attention to the knowledge contained in the data. In [29], an efficient reduction technique has been proposed; it is based on density-based clustering. Each cluster is represented by a set of carefully selected data-points, called representatives. However, selecting representatives is still a challenge in terms of quality and size [27,28].

The best way to represent a spatial cluster is by its shape and density. The shape of a cluster is represented by its boundary points (called contour) (see Fig. 1). Many algorithms for extracting the boundaries from a cluster can be found in the literature [10,16,18,30,31]. We use an algorithm based on triangulation to generate the clusters' boundaries [15]. It is an efficient algorithm for constructing non-convex boundaries. It is able to accurately characterise the shape of a wide range of different point distributions and densities with a reasonable complexity of $\mathcal{O}(n \log n)$.

3.2   Global Models

The global clusters are generated in the second phase of the DDC. This phase is also executed in a distributed fashion but, unlike the first phase, it has communications overheads. This phase consists of two main steps, which can be repeated until all the global clusters were generated. First, each leader collects the local clusters of its neighbours. Second, the leaders will merge the local clusters using the overlay technique. The process of merging clusters will continue until we reach the root node. The root node will contain the global clusters (see Fig. 1).

In DDC we only exchange the boundaries of the clusters. The communications can be synchronous or asynchronous. We implemented this phase using both types of communications. An evaluation model is presented in the next Section.

The pseudo codes of the two phases of the DDC framework are given in Algorithms 1 and 2.

Algorithm. DDC algorithm: Phase 1: Local clustering

Initialisation
$Node_i \in N$, N: The total nodes in the system.
input : X_i: Dataset Fragment, $Params_i$: Input parameters for the local
          clustering: $Params_i = (Eps_i, MinPts_i)$ for DBSCAN for example
output: C_i: Cluster's contours of $Node_i$

foreach $Node_i$ **do**
  $L_i = \text{Local\_Clustering}(X_i, Params_i)$;
  // $Node_i$ executes a clustering algorithm locally.
  $C_i = \text{Contour}(L_i)$;
  // $Node_i$ executes a contour algorithm locally.
return (C_i);

---

4   DDC Evaluation

In order to evaluate the performance of the DDC approach, we use different local clustering algorithms. For instance, with both K-Means [6] and DBSCAN [7,8], the DDC approach outperforms existing algorithms in both quality of its results and response time including K-Means and DBSCAN applied to the whole dataset. In this section we evaluate its speed-up, scalability, and which architecture is more appropriate to implement it. In addition, we compare DDC with the sequential version of the basic clustering algorithm used within DDC.

The proposed approach is more developed for distributed systems than pure parallel systems. Therefore, it is worth analysing the benefits of using synchronous or asynchronous processing mechanism, as distributed systems are asynchronous and the blocking operations have a strong impact in communication time [33].

Algorithm. DDC algorithm: Phase 2: Merging

---

input : D: Tree degree, C_i: Local cluster's contours generated by $Node_i$ in the phase 1

output: $C_{G_{k,level}}$: Global Cluster's contours (global results, level=0)

repeat
     $level = treeHeight$;
     $Node_i$ joins a group $G_{K,Level}$ of D elements;
     `// `$Node_i$` joins its neighbourhood`
     $Node_j$=ElectLeaderNode($G_{K,Level}$);
     `// `$Node_j$` is the leader of the group G`
     `// In parallel`
     foreach $Node_i \in G_{K,Level}$ **do**
         if $(i <> j)$ **then**
             Send $(C_i, Node_j)$;
             `// Each node sends its contours to others nodes in the`
                `same group of neighbourhood`
         else
             Recv $(C \equiv (\{C_l\}, Node_l))$;
             `// If the node is the leader, it will receive the others`
                `node's contours in the same group of neighbourhood`
             $G_{K,Level}$= Merge (C_i, C);
             `// Merge the overlapping contours`
     $level$ - - ;
until *(level == 0)*;
return $(C_{G_{k,0}})$

In synchronous model, as illustrated in Fig. 2, although machines M_3 and M_4 have finished their computations before M_1 and M_2, they can not send their results until M_1 and M_2 finish as well. In this model, Not only the computations and communications are not overlapped but also the machines which finished early wasted sometime waiting for the other to finish [33].

In asynchronous model the machines which finished early can advance to the next step. The machines manage their communications and the 1st and 2nd phase overlap. This model is much more suitable for distributed computing, where the nodes are heterogeneous and the communications are usually slow. As can be seen in the example given in Fig. 3, M_3 and M_4 start merging their results before M_1 and M_2 finish their computations.

4.1 DDC Computational Complexity

Let M be the number of nodes and n_i the dataset given to each node v_i in the system. The complexity of our approach is the sum of its components' complexity; local mining, local reduction, and global aggregation.

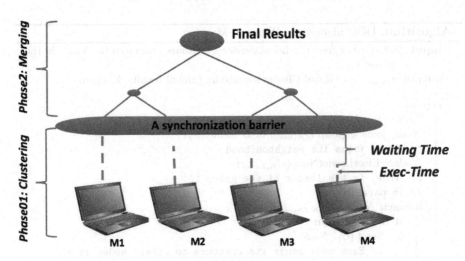

Fig. 2. Synchronous communications.

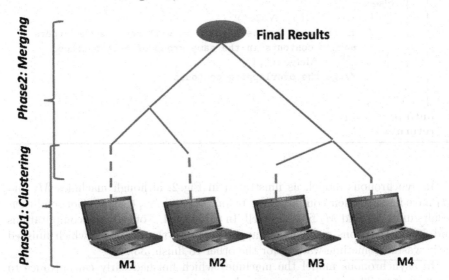

Fig. 3. Asynchronous communications.

Phase1 - Local Clustering: Let $\Gamma(n_i)$ denote the local clustering algorithm running on node (v_i), and $\Delta(c_i)$ be the time required to execute the reduction algorithm. The cost of this phase is given by:

$$T_{Phase_1} = \operatorname*{Max}_{i=1}^{M}(\Gamma(n_i) + \Delta(c_i)) \tag{1}$$

Where c_i is the cluster points generated by node v_i. Note that the reduction algorithm is of complexity $\mathcal{O}(c_i \log c_i)$.

Phase2 - Aggregation: The aggregation depends on the hierarchical combination of contours of local clusters. As the combination is based on the intersection of edges from the contours, the complexity of this phase is $\mathcal{O}(w_i \log w_i + p)$. Where w_i is the total vertices of the contours by node v_i and p is the intersection points between edges of different contours (polygons).

Total Complexity: The total complexity of the DDC approach, assuming that the local clustering algorithm is DBSCAN which is of complexity $\mathcal{O}(n^2)$, is:

$$T_{Total} = \mathcal{O}(n_i^2) + \mathcal{O}(c_i \log c_i) + \mathcal{O}(w_i \log w_i + p) \simeq \mathcal{O}(n_i^2) \qquad (2)$$

4.2 DDC Speedup

The DDC speedup is calculated against the sequential version of the approach. The sequential version consists clustering all the data on one machine. Therefore, it does require neither reduction nor aggregation. Let T_1 be the execution time of the sequential version and T_p the execution time of the DDC on p nodes. The speedup α is given by

$$\alpha = \frac{T_1}{T_p} \qquad (3)$$

Note that if the complexity of the clustering algorithm is polynomial then the optimal speedup that can be reached is P, under the condition that there is no overhead due to communications and extra work. If the complexity of the clustering algorithm is $\mathcal{O}(n^2)$ then the optimal speedup can be P^2; this is called *super speedup*. In the following section we will evaluate the speedup in the case of DBSCAN.

5 Experimental Results

We have implemented our approach on a distributed computing system. The distributed computing system consists of heterogeneous desktops (different CPUs, OSs, memory sizes, loads, etc.). We use JADE (Java Agent DEvelopment), as a development platform to implement the approach. JADE is based on a 2P2 communication architecture, It allows to use heterogeneous processing nodes, it is scalable, and dynamic [5,13].

The system nodes (desktops) are connected to local area networks. This allows us to add as many nodes as required, depending on the experiment. Table 1 lists types of machines used to perform the experiments. The main goal here is to demonstrate the performance of the DDC in a heterogeneous distributed computing environment.

We used two benchmarks of datasets from Chameleon [19]. These are commonly used to test and evaluate clustering. Table 2 gives details about the datasets.

The DDC approach is tested using various partitions of different sizes. Various scenarios were created based on the goals of the experiments. These scenarios

Table 1. The characteristics of the used machines

Machine's name	Operating system	Processor	Memory
Dell-XPS L421X	Ubuntu (V.14.04 LTS)	1.8 GHz*4 Intel Core i5	8 GB
Dell-Inspiron-3721	Ubuntu (V.14.04 LTS)	2.00 GHz*4 Intel Core i5	4 GB
Dell-Inspiron-3521	Ubuntu (V.16.04 LTS)	1.8 GHz*4 Intel Core i5	6 GB
iMac-Early 2010	cinux Mint (V.17.1 Rebecca)	3.06 GHz*2	4 GB
Dell-Inspiron-5559	Ubuntu (V.16.04 LTS)	2.30 GHz*4 Intel Core i5	8 GB
iMac-Early 2009	OS X El capitan (V.10.11.6)	2.93 *2 GHz Intel Core Due	8 GB
MacBook air	OS X El capitan (V.10.11.3)	1.6 *2 GHz Intel Core i5	8 GB

Table 2. Datasets

Benchmark	Size	Descriptions
D1	10,000 points	Different shapes, with some clusters surrounded by others
D2	30,000 points	2 small circles, 1 big circle and 2 linked ovals

mainly differ on the way the datasets are divided among the processing nodes of the distributed platform. For each scenario, we recorded the execution time for the local clustering, the merging step including (contour calculations, aggregation time and Idle time). Finally we capture also the total execution time that the approach takes to finish all the steps. in the following we describe the different scenarios considered.

5.1 Experiment I

In this scenario we give each machine a random chunk of the dataset, the size of the partition that was generated for each machine is in the range between 1500 points and 10000 points. As the dataset is relatively small we chose eight machines for the computing platform.

Table 3 shows the execution time taken by each machine to run the algorithm (step one and step two) using synchronous and asynchronous communications respectively, it also shows the overall time taken to finish all the steps.

From Table 3, we can see that the time taken by each machine to accomplish the first step of the algorithm is the same for both synchronous and asynchronous, whereas the time of the second step is different. We can also notice that each machine returns different execution time of the whole algorithm. This is because the machines have different capacities (see Table 1).

The total execution time of the algorithm while using synchronous communication is smaller compared to when using synchronous communication. This is because in synchronous communications, machines have more waiting time (up to 60% waiting time).

Table 3. Time (ms) taken by eight machines to run scenario I using synchronous and asynchronous communications

Machine	DS size	Synchronous			Asynchronous		
		STEP1	STEP2	Time	STEP1	STEP2	Time
M1	10000	21270	1104	22374	21270	554	21824
M2	2500	1060	20862	21922	1060	2515	3575
M3	3275	5093	16930	22023	5093	2017	7110
M4	5000	4592	17644	22236	4591	2620	7211
M5	1666	227	21642	21869	227	391	618
M6	2000	292	21736	22028	292	416	708
M7	5000	7520	14665	22185	7515	13949	21464
M8	1500	200	21842	22042	195	4605	4800
		Total exec-time		22374	Total exec-time		21824

5.2 Experiment II

In this scenario we allocate the whole dataset size to one machine and the remaining machines were allocated one eight of the dataset each. This scenario is chosen to show the worst case of waiting time.

Table 4 shows the execution time taken by each machine to execute the DDC technique (step one and step two) using synchronous and asynchronous communications respectively, it also shows the overall time taken to finish all the steps.

From Table 4, we can notice that the difference between the execution times of the synchronous and asynchronous DDC is still significant. Because with synchronous communications the machines need to wait for the last machine to finish its first step before they all start merging their results (step 2), whereas for asynchronous model the seven machines did the merging (step2) while the last machine finishes its clustering (step1).

5.3 Experiment III

In this scenario we allocate to seven machines the whole dataset and the one machine was allocated one eight of the dataset. This scenario is chosen to show the effect of the complexity of the local clustering complexity on the machines and on the waiting time of some powerful machines.

Table 5 shows the execution time taken by each machine to run the algorithm (step one and step two) using synchronous and asynchronous communications respectively, it also shows the overall time taken to finish all the steps.

This scenario is the opposite of the previous scenario. Unlike the previous scenarios, Table 5 shows that the difference between the execution times of synchronous and asynchronous versions of the DDC is smaller. This is because in

Table 4. Time (ms) taken by eight machines to run scenario II using synchronous and asynchronous communications

Machine	DS size	Synchronous			Asynchronous		
		STEP1	STEP2	Time	STEP1	STEP2	Time
M1	10000	21270	973	22243	21270	595	21865
M2	1250	215	21775	21990	215	518	733
M3	1250	640	21383	22023	640	20100	20740
M4	1250	304	21730	22034	304	497	801
M5	1250	161	22034	22195	161	394	555
M6	1250	171	21856	22027	170	286	456
M7	1250	245	21918	22163	245	509	754
M8	1250	185	21854	22039	185	858	1043
		Total exec-time		22243	Total exec-time		21865

both cases the machines spend more time finishing the first step, therefore, the waiting time is less for synchronous over asynchronous model.

Table 5. Time (ms) taken by eight machines to run scenario III using synchronous and asynchronous communications

Machine	DS size	Synchronous			Asynchronous		
		STEP1	STEP2	Time	STEP1	STEP2	Time
M1	10000	21270	35978	57248	21270	905	22175
M2	10000	21590	34869	56459	21590	11513	33103
M3	10000	53005	3008	56013	53005	3292	56297
M4	10000	32424	24691	57115	32424	6996	39420
M5	10000	17364	38493	55857	17364	4612	21976
M6	10000	15841	41237	57078	15841	2066	17907
M7	10000	38732	18483	57215	38727	18459	57186
M8	1250	185	56915	57100	184	16077	16261
		Total exec-time		57248	Total exec-time		57186

5.4 Experiment IV

In this scenario we took into account the machines capabilities and we divide the datasets according to their capacities. Therefore the work load is evenly distributed among them and we expect them to finish the first phase more or less at the same time. This allows to reduce the waiting time of the machines and follow immediately with the second phase. The total execution times of synchronous and asynchronous versions should be the same. This case favours more the synchronous implementation of the approach.

As predicted, Table 6 shows that there is no significant difference between the two execution times. Note that the little difference in favour of the synchronous version is due to the fact that in the asynchronous model the machines still need to execute the algorithm that checks which one finished first and receive the contours for merging.

Table 6. Time (ms) taken by eight machines to run scenario IV using synchronous and asynchronous communications

Machine	DS size	Synchronous			Asynchronous		
		STEP1	STEP2	Time	STEP1	STEP2	Time
M1	1500	256	1505	1761	256	1159	1415
M2	1660	260	598	858	260	1512	1772
M3	500	252	1061	1313	252	626	878
M4	1000	253	621	874	253	608	861
M5	1500	255	1492	1747	255	600	855
M6	1400	260	605	865	260	514	774
M7	1000	259	1030	1289	259	939	1198
M8	1500	250	603	853	250	1500	1750
		Total exec-time		1761	Total exec-time		1772

5.5 Effective Speedup

The goal here is to compare our parallel clustering to the sequential algorithm and show the DDC speedup over the sequential version of clustering, as mentioned in Eq. 3.

Considering the best scenario of executing the sequential version of DBSCAN on the fastest machine in the system. For instance, $T_1 = 15841$ ms. Clustering a partition of the same dataset on the same machine will take $= T_1^d = 258$ ms. The execution time of the DDC on the same datasets on eight heterogeneous machines with load balancing is $T_p = 1761$ ms (see Table 6). Therefore, from Eq. 3, we can deduce a speedup of 9, which is still a super-linear speedup. In the next section we will show how many processing nodes are required to cluster a dataset of size N.

5.6 Scalability

The goal here is to show that the DDC technique scales well and also we can dynamically determine the optimal number of processing nodes required to cluster a dataset of size N. We consider two datasets, the first dataset D_1 contains 10,000 data points and the second D_1 contains 30,000 data points. Figure 4 shows the execution time (y_axis is in log_2) against the number of machines in the system using the first dataset and Fig. 5 shows the execution time (y_axis is in log_2) against the number of machines in the system using the second dataset contains 30,000 data points.

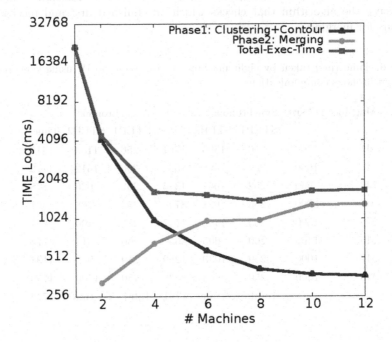

Fig. 4. Scalability experiment using dataset T_1.

As one can see, from both Figs. 4 and 5, the execution time of the first phase (Clustering and Contour) keeps decreasing as the number of machines in the distributed system increases. However, the time of the second phase (merging) keeps increasing gradually with the number of machines in the distributed system that is because the amount of communications in the second phase increases when the number of machines increases.

In addition, the total execution time of the algorithm (which is the sum of the two times, phase one and two) keep decreasing as the number of processing nodes increases until it reaches a certain points where the total execution time starts to increase (at 8 machines for dataset D_1 and at 16 machines for dataset D_2). The optimal number of processing nodes required to execute DDC is returned when the overhead of the approach exceeds the execution time of the local clustering. This is a very interesting characteristic, as one can determine the number of machines that can be allocated in advance.

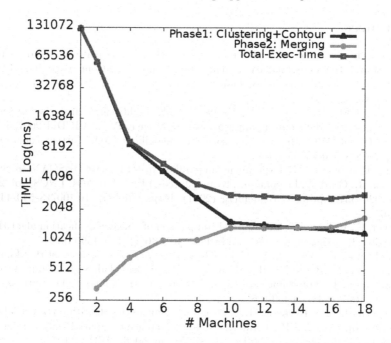

Fig. 5. Scalability experiment using dataset T_2.

6   Conclusion

In this paper, we proposed an efficient and flexible distributed clustering framework that can work with existing data mining algorithms. The approach exploits the processing power of the distributed platform by maximising the parallelism and minimising the communications and mainly the size of the data that is exchanged between the nodes of the system. It is implemented using both synchronous and asynchronous communications, and the results were significantly in favour of the asynchronous model. The approach has an efficient data reduction phase which reduces significantly the size of the data exchanged therefore, it deals with the problem of communication overhead. The DDC approach has a super-linear speedup when the complexity of the local clustering has an NP complexity. We also can determine the optimal number of processing nodes in advance.

Acknowledgement. The research work is conducted in the Insight Centre for Data Analytics, which is supported by Science Foundation Ireland under Grant Number SFI/12/RC/2289.

References

1. Aouad, L., Le-Khac, N.A., Kechadi, T.: Image analysis platform for data management in the meteorological domain. In: 7th Industrial Conference, ICDM 2007, Leipzig, Germany, July 14-18, 2007. Proceedings. vol. 4597, pp. 120–134. Springer, Heidelberg (2007)
2. Aouad, L.M., Le-Khac, N.-A., Kechadi, T.M.: Lightweight clustering technique for distributed data mining applications. In: Perner, P. (ed.) ICDM 2007. LNCS (LNAI), vol. 4597, pp. 120–134. Springer, Heidelberg (2007). https://doi.org/10.1007/978-3-540-73435-2_10
3. Arlia, D., Coppola, M.: Experiments in parallel clustering with DBSCAN. In: Sakellariou, R., Gurd, J., Freeman, L., Keane, J. (eds.) Euro-Par 2001. LNCS, vol. 2150, pp. 326–331. Springer, Heidelberg (2001). https://doi.org/10.1007/3-540-44681-8_46
4. Bauer, E., Kohavi, R.: An empirical comparison of voting classification algorithms: bagging, boosting, and variants. Mach. Learn. **36**(1), 105–139 (1999)
5. Bellifemine, F., Bergenti, F., Caire, G., Poggi, A.: Jade-a java agent development framework. In: Bordini, R.H., Dastani, M., Dix, J., El Fallah Seghrouchni, A. (eds.) Multi-agent Programming, pp. 125–147. Springer, Heidelberg (2005). https://doi.org/10.1007/0-387-26350-0_5
6. Bendechache, M., Kechadi, M.T.: Distributed clustering algorithm for spatial data mining. In: 2015 2nd IEEE International Conference on Spatial Data Mining and Geographical Knowledge Services (ICSDM), pp. 60–65. IEEE (2015)
7. Bendechache, M., Kechadi, M.T., Le-Khac, N.A.: Efficient large scale clustering based on data partitioning. In: 2016 IEEE International Conference on Data Science and Advanced Analytics (DSAA), pp. 612–621. IEEE (2016)
8. Bendechache, M., Le-Khac, N.A., Kechadi, M.T.: Hierarchical aggregation approach for distributed clustering of spatial datasets. In: 2016 IEEE 16th International Conference on Data Mining Workshops (ICDMW), pp. 1098–1103. IEEE (2016)
9. Brecheisen, S., Kriegel, H.-P., Pfeifle, M.: Parallel density-based clustering of complex objects. In: Ng, W.-K., Kitsuregawa, M., Li, J., Chang, K. (eds.) PAKDD 2006. LNCS (LNAI), vol. 3918, pp. 179–188. Springer, Heidelberg (2006). https://doi.org/10.1007/11731139_22
10. Chaudhuri, A., Chaudhuri, B., Parui, S.: A novel approach to computation of the shape of a dot pattern and extraction of its perceptual border. Comput. Vis. Image Understranding **68**, 257–275 (1997)
11. Chen, M., Gao, X., Li, H.: Parallel DBSCAN with priority r-tree. In: 2010 The 2nd IEEE International Conference on Information Management and Engineering (ICIME), pp. 508–511. IEEE (2010)
12. Coppola, M., Vanneschi, M.: High-performance data mining with skeleton-based structured parallel programming. Parallel Comput. **28**(5), 793–813 (2002)
13. Cortese, E.: Benchmark on jade message transport system (2005). http://jade.cselt.it/doc/tutorials/benchmark/JADERTTBenchmark.htm
14. Dhillon, I.S., Modha, D.S.: A data-clustering algorithm on distributed memory multiprocessors. In: Zaki, M.J., Ho, C.-T. (eds.) LSPDM 1999. LNCS (LNAI), vol. 1759, pp. 245–260. Springer, Heidelberg (2002). https://doi.org/10.1007/3-540-46502-2_13
15. Duckhama, M., Kulikb, L., Worboysc, M., Galtond, A.: Efficient generation of simple polygons for characterizing the shape of a set of points in the plane. Pattern Recogn. **41**, 3224–3236 (2008)

16. Edelsbrunner, H., Kirkpatrick, D.G., Seidel, R.: On the shape of a set of points in the plane. IEEE Trans. Inf. Theory **29**(4), 551–559 (1983)
17. Ester, M., Kriegel, H.P., Sander, J., Xu, X.: A density-based algorithm for discovering clusters in large spatial databases with noise. KDD **96**, 226–231 (1996)
18. Fadilia, M., Melkemib, M., ElMoataza, A.: Pattern Recognition Letters: Nonconvex Onion-peeling Using a Shape Hull Algorithm, vol. 24. Elsevier, Amsterdam (2004)
19. Fränti, P.: Clustering datasets (2015). http://cs.uef.fi/sipu/datasets/
20. Fu, Y.X., Zhao, W.Z., Ma, H.F.: Research on parallel DBSCAN algorithm design based on MapReduce. In: Advanced Materials Research. vol. 301, pp. 1133–1138. Trans Tech Publications (2011)
21. Garg, A., Mangla, A., Bhatnagar, V., Gupta, N.: PBIRCH: a scalable parallel clustering algorithm for incremental data. In: 10th International Symposium on Database Engineering and Applications (IDEAS-2006), pp. 315–316 (2006)
22. Geng, H., Deng, X., Ali, H.: A new clustering algorithm using message passing and its applications in analyzing microarray data. In: Proceedings of Fourth International Conference on Machine Learning and Applications, pp. 6–pp. IEEE (2005)
23. Ghosh, S.: Distributed Systems: An Algorithmic Approach. CRC Press, Boca Raton (2014)
24. Guo, Y., Grossman, R.: A fast parallel clustering algorithm for large spatial databases, high performance data mining. Data Mining Knowl. Discov. (2002)
25. Han, J., Kamber, M., Pei, J.: Data Mining: Concepts and Techniques, 3rd edn, pp. 1–38. Morgan Kaufmann Publishers Inc., San Francisco (2011). ISBN 0123814790, ISBN 9780123814791
26. Jain, A.K., Murty, M.N., Flynn, P.J.: Data clustering: a review. ACM Comput. Surv. (CSUR) **31**(3), 264–323 (1999)
27. Januzaj, E., Kriegel, H.-P., Pfeifle, M.: DBDC: density based distributed clustering. In: Bertino, E., Christodoulakis, S., Plexousakis, D., Christophides, V., Koubarakis, M., Böhm, K., Ferrari, E. (eds.) EDBT 2004. LNCS, vol. 2992, pp. 88–105. Springer, Heidelberg (2004). https://doi.org/10.1007/978-3-540-24741-8_7
28. Laloux, J.F., Le-Khac, N.A., Kechadi, M.T.: Efficient distributed approach for density-based clustering. In: 20th IEEE International Workshops on Enabling Technologies: Infrastructure for Collaborative Enterprises (WETICE), pp. 145–150, 27–29 June 2011
29. Le-Khac, N.A., Bue, M., Whelan, M., Kechadi, M.-T.: A knowledge based data reduction for very large spatio-temporal datasets. In: International Conference on Advanced Data Mining and Applications (ADMA 2010), 19–21 November 2010
30. Melkemi, M., Djebali, M.: Computing the shape of a planar points set. Elsevier Sci. **33**, 1423–1436 (2000)
31. Moreira, A., Santos, M.Y.: Concave hull: a k-nearest neighbours approach for the computation of the region occupied by a set of points. In: International Conference on Computer Graphics Theory and Applications (GRAPP-2007), Barcelona, Spain, pp. 61–68, 8–11 March 2007
32. Rokach, L., Schclar, A., Itach, E.: Ensemble methods for multi-label classification. Expert Syst. Appl. **41**, 7507–7523 (2014)
33. Solar, R., Borges, F., Suppi, R., Luque, E.: Improving communication patterns for distributed cluster-based individual-oriented fish school simulations. Procedia Comput. Sci. **18**, 702–711 (2013)
34. Wu, X., Zhu, X., Wu, G.Q., Ding, W.: Data mining with big data. IEEE Trans. Knowl. Data Eng. **26**(1), 97–107 (2014)

35. Xu, X., Jäger, J., Kriegel, H.P.: A fast parallel clustering algorithm for large spatial databases. Data Mining Knowl. Discov. Arch. **3**, 263–290 (1999)
36. Zaki, M.J.: Parallel and distributed data mining: an introduction. In: Zaki, M.J., Ho, C.-T. (eds.) LSPDM 1999. LNCS (LNAI), vol. 1759, pp. 1–23. Springer, Heidelberg (2000). https://doi.org/10.1007/3-540-46502-2_1
37. Zhang, T., Ramakrishnan, R., Livny, M.: BIRCH: an efficient data clustering method for very large databases. In: ACM SIGMOD Record, vol. 25, pp. 103–114. ACM (1996)
38. Zhou, A., Zhou, S., Cao, J., Fan, Y., Hu, Y.: Approaches for scaling DBSCAN algorithm to large spatial databases. J. Comput. Sci. Technol. **15**(6), 509–526 (2000)

Patched Completed Local Binary Pattern is an Effective Method for Neuroblastoma Histological Image Classification

Soheila Gheisari[1]([✉]), Daniel R. Catchpoole[2], Amanda Charlton[3], and Paul J. Kennedy[1]

[1] Faculty of Engineering and Information Technology,
Centre for Artificial Intelligence, University of Technology Sydney,
Ultimo, NSW 2007, Australia
soheila.gheisari@student.uts.edu.au, paul.kennedy@uts.edu.au
[2] Biospecimens Research and Tumour Bank, Children's Cancer Research Unit,
The Kids Research Institute, The Children's Hospital at Westmead,
Westmead, NSW 2145, Australia
daniel.catchpoole@health.nsw.gov.au
[3] The Children's Hospital at Westmead, Cnr Hawkesbury Road and Hainsworth
Street, Westmead, NSW 2145, Australia
amandac@rcpa.edu.au

Abstract. Neuroblastoma is the most common extra cranial solid tumour in children. The histology of neuroblastoma has high intra-class variation, which misleads existing computer-aided histological image classification methods that use global features. To tackle this problem, we propose a new Patched Completed Local Binary Pattern (PCLBP) method combining Sign Binary Pattern (SBP) and Magnitude Binary Pattern (MBP) within local patches to build feature vectors which are classified by k-Nearest Neighbor (k-NN) and Support Vector Machine (SVM) classifiers. The advantage of our method is extracting local features which are more robust to intra-class variation compared to global ones. We gathered a database of 1043 histologic images of neuroblastic tumours classified into five subtypes. Our experiments show the proposed method improves the weighted average F-measure by 1.89% and 0.81% with k-NN and SVM classifiers, respectively.

Keywords: Neuroblastic tumour · Neuroblastoma · Classification
Binary pattern · Local patch · Image analysis
Computer-Aided Diagnosis (CAD)

1 Introduction

Neuroblastoma is the most common extra cranial solid tumour in children less than five years of age. More than 15% of childhood cancer deaths are the result of neuroblastic tumours [19]. Optimal management of neuroblastic tumours

© Springer Nature Singapore Pte Ltd. 2018
Y. L. Boo et al. (Eds.): AusDM 2017, CCIS 845, pp. 57–71, 2018.
https://doi.org/10.1007/978-981-13-0292-3_4

depends on many factors, including histopathological classification. Histological classification is performed by a medical laboratory doctor who diagnoses tumours by examining thin slices of tissue on a glass slide using an optical microscope. Pathologists commonly use the Shimada system [23] which identifies six morphologic categories of neuroblastic tumour. Computer image analysis of tumours has been shown to improve diagnostic efficiency and consistency [11], and identify previously unrecognized image features that predict prognosis [28]. Existing methods of computer-aided diagnosis (CAD) for classification of histological images, thin slices of tissue mounted on a glass slide and viewed with a microscope, are divided into two categories: segmentation-based methods and feature-based methods. Segmentation-based methods rely on morphological features such as symmetry. Feature-based methods try to extract mathematical features from the histological images and classify them without segmentation [3].

There are several factors that hinder the classification of neuroblastoma histological images based on segmentation methods. First of all, different cells in neuroblastoma histopathological images have variations in illumination. Second, they have different shapes within the same classification group which show the high intra-class variation of neuroblastoma. In general, classification methods using segmentation may fail to detect nuclei and the different cells in images because they segment different cells based on illumination and shape. All existing feature-based classification methods used global features, extracted from the whole image, which are sensitive to intra-class variations. To our best knowledge, there is no feature-based method to address the intra-class variation problem in neuroblastoma histological image classification.

This paper proposes a local feature extraction method for classification of neuroblastoma histological images to tackle the intra-class variation problem. The contributions of this paper are:

1. We develop Patched Completed Local Binary Pattern (PCLBP) based on Completed Local Binary Pattern (CLBP) [9] for classification of neuroblastoma histological images with intra-class variation.
2. We apply PCLBP on neuroblastoma histological images with a complex histology to classify them into five different categories.
3. We evaluate our method by comparing with a state-of-the-art benchmark which shows the effectiveness of our method in the classification of neuroblastoma histological images.

The rest of this paper is as follows. Section 2 presents related work. Section 3 describes the gathered dataset. Section 4 presents the proposed Patch Completed Local Binary Pattern. Section 5 shows the experimental results. Section 6 presents the discussion and finally Sect. 7 concludes the work.

2   Related Work

Most proposed techniques for classification of histological images rely on morphological features [27]. However, variability of illumination and appearance of

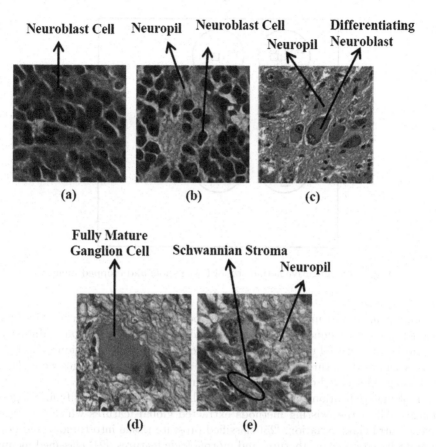

Fig. 1. Neuroblastic tumour categories: (a) undifferentiated neuroblastoma, (b) poorly-differentiated neuroblastoma, (c) differentiating neuroblastoma, (d) ganglioneuroma, and (e) ganglioneuroblastoma

different cells in the images makes classification based on segmentation more challenging. [13] proposed an automated system for segmentation and classification of breast cancer's nuclei. [15] segmented blood cell images using watershed technique to identify chronic lymphocytic leukemia by extraction of nucleus and cytoplasm mask. [6] classified breast cancer images using nuclei shape and size. [16] classified nuclei into normal and cancer based on appearance and identified the cancer glands in prostate cancer. [22] segmented nuclei and classified with AdaBoost based on intensity and morphological features of nuclei.

[12] classified neuroblastoma into three categories: undifferentiated, poorly-differentiated, and differentiating. They segmented the images at each resolution level into cellular, neuropil, and background elements and classified neuroblastoma histological images by integrating classifiers such as Linear Discriminative Analysis (LDA) [14], Support Vector Machine (SVM) [5], and k-Nearest Neighbor (k-NN) [7]. [26] proposed a four stage algorithm to classify neuroblastoma

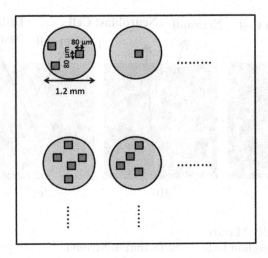

Fig. 2. Quantitative actual size of tissue spots and cropped images

tumour images into undifferentiated and poorly-differentiated using the Otsu segmentation technique [18]. The whole slide image was partitioned. Intensity variation was reduced using an image enhancement technique, regions of interest were segmented by thresholding techniques and histological images were classified using rules based on the Shimada scheme.

In some applications, feature extraction has been used to classify histological images. All of the existing methods extracted global features which are sensitive to intra-class variation. [25] classified prostate tissue into tumour and nontumour based on color, texture, and morphologic features. [21] classified meningioma using wavelet package transform and Local Binary Patterns [17]. [29] used a combination of curvelet transform, gray level co-occurrence matrix, and the Completed Local Binary Patterns (CLBP) as features to stratify breast cancer tumours from histological images. [8] used texture features to identify the glandular elements within images of prostate tissue. They applied k-means to cluster the image components. [24] applied a completed modeling of the Local Binary Pattern (LBP), based on three components extracted from the 8-neighborhood: center pixel, sign, and magnitude. The center pixel is coded into a binary bit after global thresholding. The difference signs and magnitudes are coded in binary format so that they can be combined to form the final Completed Local Binary Pattern histograms. [24] used the combination of Completed Local Binary Patterns with k-NN and SVM to classify a large dataset of breast cancer histological images into benign and malignant classes. Classification of neuroblastoma histological images remains challenging due to the intra-class variation.

3   Dataset

There is a lack of large and publicly available image datasets for analysis of neuroblastic tumours, which significantly hinders development and validation of methods. Therefore, we gathered a dataset of images from neuroblastic tumours from The Tumour Bank of the Kid's Research Institute at The Children's Hospital at Westmead, Sydney, Australia. Tumour access is compliant with local policy, national legislation, and ethical mandates to use the human tissue in research. All patient specific details were removed and a de-identified dataset was used for this research. The initial dataset consisted of images of tissue microarrays (TMA) of neuroblastic tumours, scanned by the Aperio ScanScope system. Each slide was composed of 20 to 40 1.2 mm cores of neuroblastic tumour, stained with haematoxylin and eosin (H&E) and cut at 3 μm. In this method, the contrast between different cells which have different colors is increased. Staining with H&E allows observation of histological structures. TMA images were in svs format with resolution 0.2 μm, images were viewed and extracted using ImageScope software [1]. Tissue cores were classified by experts into five different categories: poorly-differentiated, differentiating, undifferentiated, ganglioneuroma, and ganglioneuroblastoma, according to the Shimada classification system. Representative images in the categories are shown in Fig. 1.

Areas best representative of each category, and devoid of artefacts, were selected from each tissue core by an expert histopathologist [AC]. At 40× magnification, cropped image size was 300 × 300 pixels with real size 80 ×80 μm which

Table 1. Number of different categories of neuroblastic tumour cropped images

Category of neuroblastic tumour	Number of cropped images	Number of patients
Poorly-differentiated	571	77
Differentiating	187	12
Undifferentiated	155	10
Ganglioneuroma	84	18
Ganglioneuroblastoma	46	8
Total	1043	125

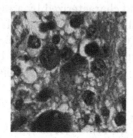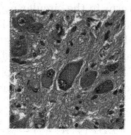

Fig. 3. An example of high intra-class variation of differentiating neuroblastoma

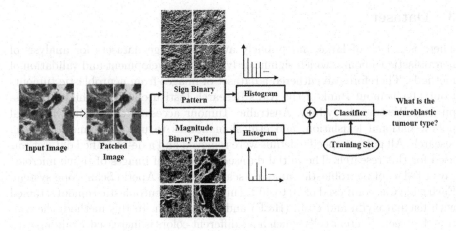

Fig. 4. The scheme of the proposed method

is approximately one third of the area of an optical microscope high power field of view. Figure 2 shows the quantitative actual size of tissue spots and cropped images.

This size was chosen as a compromise between being large enough to capture diagnostic features of each category and small enough for computational cost. Numbers of images in our dataset are given in Table 1. It is much larger in terms of patients and images than the datasets used by [12] and [26]. Moreover, the intra-class variation of neuroblastoma cells in the gathered dataset is very high which means different cells in neuroblastoma histological images within the same patients in the same class have different shapes. An example of the high intra-class variation of differentiating neuroblastoma is shown in Fig. 3. As can be seen, both of them are differentiating type but their cells have different shapes.

4   Patch Completed Local Binary Pattern (PCLBP)

Before describing our PCLBP algorithm, we first describe the algorithm it is based on Completed Local Binary Pattern (CLBP) is one of the latest variants of Local Binary Pattern (LBP) [17]. The LBP operator computes the distribution of binary patterns in the circular neighborhood characterized by a radius R and a number of neighbors P. The idea is to threshold neighboring pixels, compared to the central pixel to the P neighbors. If the intensity of a neighbor pixel is greater than or equal to that of the central pixel the value 1 is assigned, otherwise 0. Therefore, a binary pattern is obtained from the neighborhood. The LBP function at pixel p is (from [17])

$$LBP(f(X,Y)) = \sum_{i=0}^{P-1} 2^i \cdot u(f(X_i,Y_i) - f(X,Y)) \tag{1}$$

where $f(X_i, Y_i)$ and $f(X, Y)$ are grey levels of pixels (X_i, Y_i) and (X, Y) and $u(\cdot)$ is the unit step function. The CLBP is a completed modeling of LBP [9] which is based on three components extracted from the local region: center pixel, sign, and magnitude. The center pixel is coded by a binary code after thresholding, with the threshold set as the average grey level of the whole image. For computing the sign and magnitude, a neighborhood of radius R and number of neighbors P is considered. Signs and magnitudes are computed and coded by a specific operator into the binary format so that they can be combined to form the final CLBP histograms [9].

Our approach, Patched Completed Local Binary Pattern (PCLBP), extends CLBP. The overall framework consists of four stages as shown in Fig. 4. First, the images are partitioned into equal-sized square patches. Second, Sign Binary Patterns (SBPs) and Magnitude Binary Patterns (MBPs) are computed within patches. Third, histograms of SBPs and MBPs are computed and concatenated to build a feature vector for each patch. A feature vector for the whole image is created by concatenating the feature vectors of all patches. Finally, the input image is classified by comparing the related feature vector with the feature vectors of all images in the gallery. Following we describe the algorithm in detail.

Given an $N \times N$ pixel input image, we partition it into $W \times W$ pixel non-overlapping patches. We indicate all points in the patch with p and q indices, ranging from 1 to N/W, as

$$p = \left\lfloor \frac{X}{W} \right\rfloor + 1, \quad q = \left\lfloor \frac{Y}{W} \right\rfloor + 1 \tag{2}$$

where $0 \leq X \leq N$ and $0 \leq Y \leq N$ are the coordinates of the input neuroblastoma image and $\lfloor \cdot \rfloor$ is the floor function. The (p, q)th patch in the input neuroblastoma image (see Fig. 5) is defined as

$$f_{p,q}(X^{pq}, Y^{pq}) = f(W(p - 1) + X^{pq}, W(q - 1) + Y^{pq}) \tag{3}$$

where $f(X, Y)$ denotes the original image. X^{pq}, Y^{pq} denote the coordinate of the (p, q)th patch.

The local differences of the 8-neighborhood around (X_0^{pq}, Y_0^{pq}), see Fig. 6, are computed as

$$f'_{p,q,k}(X_0^{pq}, Y_0^{pq}) = f_{p,q}(X_k^{pq}, Y_k^{pq}) - f_{p,q}(X_0^{pq}, Y_0^{pq}); k = 1, \ldots, 8 \tag{4}$$

The SBP of the (p, q)th patch, $SBP_{p,q}(f_{p,q}(X_0^{pq}, Y_0^{pq}))$, is defined as the concatenation of 8 bits as

$$SBP_{p,q}(f_{p,q}(X_0^{pq}, Y_0^{pq})) = \{u(f'_{p,q,1}(X_0^{pq}, Y_0^{pq})), \ldots, u(f'_{p,q,8}(X_0^{pq}, Y_0^{pq}))\} \tag{5}$$

where $u(x)$ is the unit step function:

$$u(x) = \begin{cases} 1 & \text{if } x \geq 0 \\ 0 & \text{if } x < 0 \end{cases} \tag{6}$$

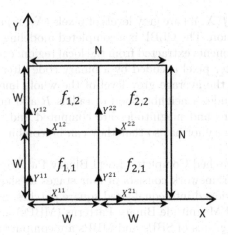

Fig. 5. Coordinates in the patched image. Here, we assume $W = N/2$

(X_1^{pq}, Y_1^{pq})	(X_2^{pq}, Y_2^{pq})	(X_3^{pq}, Y_3^{pq})
(X_8^{pq}, Y_8^{pq})	(X_0^{pq}, Y_0^{pq})	(X_4^{pq}, Y_4^{pq})
(X_7^{pq}, Y_7^{pq})	(X_6^{pq}, Y_6^{pq})	(X_5^{pq}, Y_5^{pq})

Fig. 6. An 8-neighborhood around (X_0^{pq}, Y_0^{pq})

and the MBP for the (p, q)th patch is defined as

$$MBP_{p,q}(f_{p,q}(X_0^{pq}, Y_0^{pq})) = \{u(f'_{p,q,1}(X_0^{pq}, Y_0^{pq}) - m), \dots, u(f'_{p,q,8}(X_0^{pq}, Y_0^{pq}) - m)\} \tag{7}$$

where m is a threshold to be set as the average of the absolute values of all derivatives in the neuroblastoma image.

After computing the SBP and the MBP, they are converted into decimal values as

$$DSBP_{p,q}(f_{p,q}(X_0^{pq}, Y_0^{pq})) = \sum_{l=1}^{8} 2^{l-1} SBP_{p,q}^l(f_{p,q}(X_0^{pq}, Y_0^{pq})) \tag{8}$$

$$DMBP_{p,q}(f_{p,q}(X_0^{pq}, Y_0^{pq})) = \sum_{l=1}^{8} 2^{l-1} MBP_{p,q}^l(f_{p,q}(X_0^{pq}, Y_0^{pq})) \tag{9}$$

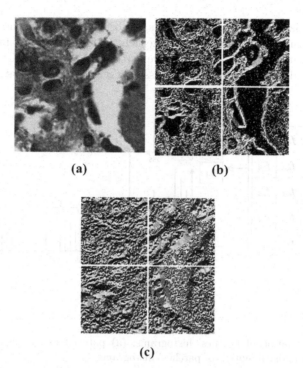

(a)   (b)

(c)

Fig. 7. An example of computed Magnitude Binary Pattern (MBP) and Sign Binary Pattern (SBP): (a) original image, (b) MBP, and (c) SBP. The optimal patch size (W) is 60 Pixels. Here, we assume $W = 150$ pixels for better visualization.

where $SBP_{p,q}^l(f_{p,q}(X_0^{pq}, Y_0^{pq}))$ and $MBP_{p,q}^l(f_{p,q}(X_0^{pq}, Y_0^{pq}))$ denotes the l-th bit of the SBP and the MBP, respectively. DSBP and DMBP for each pixel in the neuroblastoma image are computed. Figure 7 shows an example of DMBP and DSBP in an arbitrary neuroblastoma image. For each patch, we model the distribution of DSBP and DMBP using the histogram operator with 256 bins as

$$HSBP_{p,q}(f_{p,q}(X,Y)) = H\{DSBP_{p,q}(f_{p,q}(X,Y))\} \tag{10}$$

$$HMBP_{p,q}(f_{p,q}(X,Y)) = H\{DMBP_{p,q}(f_{p,q}(X,Y))\} \tag{11}$$

We concatenate the histograms of the SBP and the MBP for each patch to build a Local Histogram (LH) for each patch

$$LH_{p,q}(f_{p,q}(x,y)) = \{HSBP_{p,q}(f_{p,q}(x,y)), HMBP_{p,q}(f_{p,q}(x,y))\} \tag{12}$$

Then, we concatenate the histograms of all patches to build the Patch Completed Local Binary Pattern for the neuroblastoma image (as shown in Fig. 8)

$$PCLBP = \{LH_{p,q}(f_{p,q}(X,Y)) \,|p,q = 1, \ldots, N/W\} \tag{13}$$

Finally, two algorithms are used to classify the extracted PCLBPs: k-NN [7] and SVM [5]. We choose k-NN because of its low cost learning process and simple procedures of learning complex concepts. Also, SVM can efficiently perform a non-linear classification using different kernels to map the inputs into the high-dimensional feature spaces. Although, we use k-NN and SVM, PCLBP features can be classified using other existing classifiers without customization.

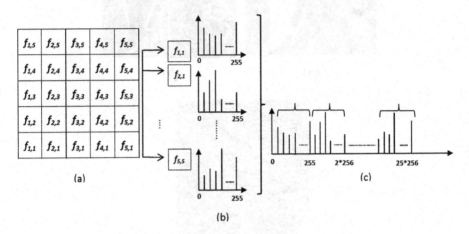

Fig. 8. Concatenation of patches' histograms: (a) patched image, (b) histogram of patches, and (c) concatenation of patches' histograms.

5 Experimental Results

In this section, we evaluate the performance of the proposed method for classification of neuroblastoma histological images. Experiments are conducted on the collected neuroblastic tumour database. The database is divided randomly into two subsets: parameter-tuning (211 images) and validation datasets (832 images). We select the optimum values for free parameters using the training dataset and fix them for the validation. Then, we evaluate the system using the validation dataset and selected parameters.

5.1 Parameter Tuning

We divide the parameter-tuning dataset into training (150 images) and testing (61 images) subsets. We train the algorithm using the training set with different parameter values, test using the testing set and compute the accuracies. To have a better estimation of the accuracy, we repeat the above procedure multiple (10) times and compute the average over all experiments.

For k-NN, the free parameters are the width of the patch (W) and the k numbers of neighbours. Accuracy was computed for k ranging from 1 to 10 and patch width $W \in \{10, 15, 30, 50, 60, 75, 100, 150, 300\}$ and the results are shown in Fig. 9. Best accuracy of 72.7% was found with $W = 60$ and $k = 5$. So we used these values in the next experiments.

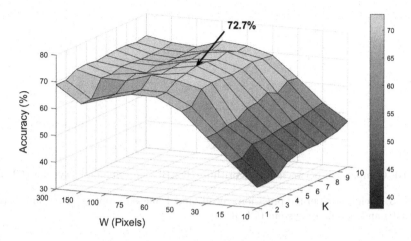

Fig. 9. Accuracy of k-NN classifier versus patch width (W) and k in parameter tuning of k-NN classifier

Table 2. Average classification accuracy of SVM over neuroblastic tumour dataset using different kernel functions

Kernel	Classification accuracy (%)
Linear	62.22
Polynomial	71.82
Radial basis function	72.33
Sigmoid	49.45

For SVM, we used the C-SVC type [2] using LIBSVM tool [4] and tested different kernels: linear, polynomial, Radial Basis Function (RBF), and sigmoid. Table 2 shows accuracies using different kernels. As can be seen, the best result is achieved using the RBF kernel, so we selected it for the next experiments. RBF parameter γ was empirically defined through experiments with best value taking $1/256$ (256 is the number of different intensities in the images).

5.2   System Validation

Here, we use the remaining 80% of the dataset which is not seen in the parameter tuning phase. It is divided into training (623 images) and validation (209 images) sets. We train the algorithm using the training set (with the parameter values selected in Sect. 5.1) and test using the validation set. We repeat this procedure multiple (10) times and report the average accuracy. Algorithm performance for k-NN and SVM is reported with the average F-measure, recall, and precision [20] weighted by number of examples in each of the five classes. Distribution of the computed F-measures for the two classifiers and feature extraction approaches over the ten trials is presented in Fig. 10 which shows that SVM works better

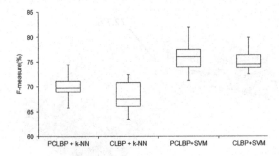

Fig. 10. Comparison between our algorithm (PCLBP) and Spanhol's system (CLBP)

Table 3. Weighted average precision, recall, and F-measure obtained by our system (PCLBP) and Spanhol's system (CLBP).

	PCLBP (%)	CLBP (%)
Precision-kNN	70.49 ± 3.37	68.22 ± 3.23
Precision-SVM	75.59 ± 3.15	74.1 ± 2.35
Recall-kNN	71.02 ± 2.87	69.53 ± 2.81
Recall-SVM	76.35 ± 3.41	76.25 ± 2.23
F-measure-kNN	70.75 ± 3.09	68.86 ± 6.04
F-measure-SVM	75.96 ± 3.27	75.15 ± 2.28

than k-NN. The t-test with P value $= 0.03$ and $\alpha = 0.05$ (significance level) shows that combination of our algorithm with SVM classifier significantly improves the accuracy of classification in comparison with k-NN classifier. Table 3 indicates that our algorithm obtains approximately 5% higher accuracy when it is combined with SVM classifier compared to the k-NN classifier. We also test the CLBP algorithm [24] on the test images as a benchmark, again reporting the weighted average of precision, recall, and F-measure. Table 3 reports the weighted average precision, recall, and F-measure of our system and Spanhol's system. The weighted average precision, recall, and F-measure of our system are better than Spanhol's system.

6   Discussion

The proposed algorithm is a new feature extraction method to classify neuroblastoma histological images into five different groups. Although a large number of methods have been proposed in the literature, our system has multiple advantages over these systems:

1. There is no feature based method to classify histological images into more than two categories. They were classified more straightforward to binary classification. However, the proposed method can classify neuroblastoma tumour images to five different categories.

2. Neuroblastoma has a complex texture with a great deal of complicated features compared to other types of cancer such as breast cancer. It is the first time that neuroblastoma histological images are classified into five different categories using a feature extraction method.
3. The proposed method extract features within small patches which are not easily detected by human eyes.

7 Conclusion

We proposed a new Patched Completed Local Binary Pattern (PCLBP) to classify neuroblastic tumours into five different categories using extracted feature vectors from histological images. The algorithm built the feature vector by extraction of SBP and MBP within local patches. The advantage of the proposed method is extraction of local features which are more robust to intra-class variation compared to global feature extraction. The evaluation was conducted on a gathered dataset with 1043 cropped images from samples of five different categories. We compare the results obtained by our system with the state-of-the-art. Results indicate that the proposed method has improved the average weighted F-measure for k-NN and SVM by 1.89% and 0.81%, respectively, compared to the benchmark. In the future work, we will apply Convolutional Deep Belief Network (CDBN) [10] to local patches to classify histological images of neuroblastic tumours.

References

1. ImageScope (2016). http://www.leicabiosystems.com/digital-pathology/digital-pathology-management/imagescope/
2. Boser, B., Guyon, I., Vapnik, V.: A training algorithm for optimal margin classifiers. In: Proceedings of the Fifth Annual Workshop on Computational Learning Theory, pp. 144–152 (1992)
3. Boucheron, L.: Object-and spatial-level quantitative analysis of multispectral histopathology images for detection and characterization of cancer. Doctoral Dissertation in University of California at Santa Barbara (2008)
4. Chang, C., Lin, C.: LIBSVM: a library for support vector machines. ACM Trans. Intell. Syst. Technol. (TIST) **2**(3), 1–27 (2011). http://www.csie.ntu.edu.tw/~cjl in/libsvm
5. Cortes, C., Vapnik, V.: Support-Vector Networks. Mach. Learn. **20**(3), 273–297 (1995)
6. Cosatto, E., Miller, M., Graf, H., Meyer, J.: Grading nuclear pleomorphism on histological micrographs. In: International Conference on Pattern Recognition, pp. 1–4 (2008)
7. Cover, T., Hart, P.: Nearest neighbor pattern classification. IEEE Trans. Inf. Theory **13**(1), 21–27 (1967)
8. Farjam, R., Soltanian-Zadeh, H., Jafari-Khouzani, K., Zoroofi, R.: An image analysis approach for automatic malignancy determination of prostate pathological images. Cytometry Part B: Clin. Cytometry **72**(4), 227–240 (2007)

9. Guo, Z., Zhang, L., Zhang, D.: A completed modeling of local binary pattern operator for texture classification. IEEE Trans. Image Process. **19**(6), 1657–1663 (2010)

10. Hinton, J., Osindero, S.: A fast learning algorithm for deep belief nets. Neural Comput. **18**(7), 1527–1554 (2006)

11. Hipp, J., Flotte, T., Monaco, J., Cheng, J., Madabhushi, A., Yagi, Y., Rodriguez-Canales, J., Emmert-Buck, M., Dugan, M., Hewitt, S., Toner, M., Tompkins, R., Lucas, D., Gilbertson, J., Balis, U.: Computer-aided diagnostic tools aim to empower rather than replace pathologists: lessons learned from computational chess. J. Pathol. Inf. **2**(1), 25 (2011)

12. Kong, J., Sertel, O., Shimada, H., Boyer, K., Saltz, J., Gurcan, M.: Computer-aided evaluation of neuroblastoma on whole-slide histology images: classifying grade of neuroblastic differentiation. Pattern Recogn. **42**(6), 1080–1092 (2009)

13. Lee, K., Street, W.: An adaptive resource-allocating network for automated detection, segmentation, and classification of breast cancer nuclei topic area: image processing and recognition. IEEE Trans. Neural Netw. **14**(3), 680–687 (2003)

14. Lehmann, E.L., Casella, G.: Theory of Point Estimation. STS. Springer, New York (1998). https://doi.org/10.1007/b98854

15. Mohammed, E., Mohamed, M., Naugler, C., Far, B.: Chronic lymphocytic leukemia cell segmentation from microscopic blood images using watershed algorithm and optimal thresholding. In: 26th IEEE Canadian Conference on Electrical and Computer Engineering, pp. 1–5 (2013)

16. Nguyen, K., Jain, A., Sabata, B.: Prostate cancer detection: fusion of cytological and textural features. J. Pathol. Inf. **2**, S3 (2011)

17. Ojala, T., Pietikainen, M., Maenpaa, T.: Multiresolution gray-scale and rotation invariant texture classification with local binary patterns. IEEE Trans. Pattern Anal. Mach. Intell. **24**(7), 971–987 (2002)

18. Otsu, N.: A threshold selection method from gray-level histograms. IEEE Trans. Syst. Man Cybern. **9**(1), 62–66 (1979)

19. Park, J., Caron, H., Eggert, A.: Neuroblastoma: biology, prognosis, and treatment. Pediatr. Clin. North Am. **55**(1), 97–120 (2008)

20. Powers, D.: Evaluation: from precision, recall and F-measure to ROC, informedness, markedness, and correlation. J. Mach. Learn. Technol. **2**(1), 37–63 (2011)

21. Qureshi, H., Sertel, O., Rajpoot, N., Wilson, R., Gurcan, M.: Adaptive discriminant wavelet packet transform and local binary patterns for meningioma subtype classification. In: Proceedings of the 11th International Conference on Medical Image Computing and Computer-Assisted Intervention, Part II. pp. 196–204 (2008)

22. Sharma, H., Zerbe, N., Heim, D., Wienert, S., Behrens, H., Hellwich, O., Hufnagl, P.: A multi-resolution approach for combining visual information using nuclei segmentation and classification in histopathological images. In: 10th International Conference on Computer Vision Theory and Applications, pp. 37–46 (2015)

23. Shimada, H., Ambros, I., Dehner, L., Hata, J., Joshi, V., Roald, B., Stram, D., Gerbing, R., Lukens, J., Matthay, K., Castleberry, R.: The international neuroblastoma pathology classification (the Shimada system). Cancer **86**(2), 364–372 (1999)

24. Spanhol, F., Oliveira, L., Caroline, P., Laurent, H.: A dataset for breast cancer histopathological image classification. IEEE Trans. Biomed. Eng. **63**(7), 1455–1462 (2016)

25. Tabesh, A., Teverovskiy, M., Pang, H., Kumar, V., Verbel, D., Kotsianti, A., Saidi, O.: Multifeature prostate cancer diagnosis and gleason grading of histological images. IEEE Trans. Med. Imaging **26**(10), 1366–1378 (2007)

26. Tafavogh, S., Meng, Q., Catchpoole, D., Kennedy, P.: Automated quantitative and qualitative analysis of the whole slide images of neuroblastoma tumour for making a prognosis decision. In: Proceedings of the IASTED 11th International Conference on Biomedical Engineering, pp. 244–251 (2014)
27. Veta, M., Pluim, J., van Diest, P., Viergever, M.: Breast cancer histopathology image analysis: a review. IEEE Trans. Biomed. Eng. **61**(5), 1400–1411 (2014)
28. Yu, K., Zhang, C., Berry, G., Altman, R., Re, C., Rubin, D., Snyder, M.: Predicting non-small cell lung cancer prognosis by fully automated microscopic pathology image features. Nat. Commun. **7**, 1–10 (2016)
29. Zhang, Y., Zhang, B., Lu, W.: Breast cancer histological image classification with multiple features and random subspace classifier ensemble. Stud. Comput. Intell. **450**, 27–42 (2013)

An Improved Naive Bayes Classifier-Based Noise Detection Technique for Classifying User Phone Call Behavior

Iqbal H. Sarker[1]($\boxtimes$), Muhammad Ashad Kabir[2], Alan Colman[1], and Jun Han[1]

[1] Department of Computer Science and Software Engineering,
School of Software and Electrical Engineering,
Swinburne University of Technology, Melbourne, Australia
{msarker,acolman,jhan}@swin.edu.au
[2] School of Computing and Mathematics, Charles Sturt University,
Sydney, NSW, Australia
akabir@csu.edu.au

Abstract. The presence of noisy instances in mobile phone data is a fundamental issue for classifying user phone call behavior (i.e., accept, reject, missed and outgoing), with many potential negative consequences. The classification accuracy may decrease and the complexity of the classifiers may increase due to the number of redundant training samples. To detect such noisy instances from a training dataset, researchers use naive Bayes classifier (NBC) as it identifies misclassified instances by taking into account independence assumption and conditional probabilities of the attributes. However, some of these misclassified instances might indicate *usages behavioral patterns* of individual mobile phone users. Existing naive Bayes classifier based noise detection techniques have not considered this issue and, thus, are lacking in *classification accuracy*. In this paper, we propose an *improved noise detection technique* based on naive Bayes classifier for effectively classifying users' phone call behaviors. In order to improve the classification accuracy, we effectively identify noisy instances from the training dataset by analyzing the behavioral patterns of individuals. We dynamically determine a *noise threshold* according to individual's unique behavioral patterns by using both the naive Bayes classifier and Laplace estimator. We use this noise threshold to identify noisy instances. To measure the effectiveness of our technique in classifying user phone call behavior, we employ the most popular classification algorithm (e.g., decision tree). Experimental results on the real phone call log dataset show that our proposed technique more accurately identifies the noisy instances from the training datasets that leads to better classification accuracy.

Keywords: Mobile data mining · Noise analysis
Naive Bayes classifier · Decision tree · Classification
Laplace estimator · Predictive analytics · Machine learning
User behavior modeling

© Springer Nature Singapore Pte Ltd. 2018
Y. L. Boo et al. (Eds.): AusDM 2017, CCIS 845, pp. 72–85, 2018.
https://doi.org/10.1007/978-981-13-0292-3_5

1   Introduction

Now a days, mobile phones have become part of our daily life. The number of mobile cellular subscriptions is almost equal to the number of people on the planet [13] and the phones are, for most of the day, with their owners as they go through their daily routines [13]. People use mobile phones for various activities such as voice communication, Internet browsing, app using, e-mail, online social network, instant messaging, etc. [13]. In recent years, researchers use various types of mobile phone data such as phone call log [12], app usages log [18], mobile phone notifications history [11], web log [8], context log [23] for different personalized applications. For instance, phone call log is used to predict users' behavior in order to build an automated call firewall or call reminder system [14].

In data mining area, classification is a function that describes and distinguishes data classes or concepts [5]. The goal of classification is to accurately classify the class labels of instances whose attribute values are known, but class values are unknown. Accurately classifying user phone call behavior from log data using machine learning techniques (e.g., decision tree) is challenging as it requires a data set free from outliers or noise [3]. However, real-world datasets may contain noise, which is anything that obscures the relationship between the features of an instance and it's behavior class [6]. Such noisy instances may reduce the classification accuracy, and increase the complexity of the classification process. It is also evident that decision trees are badly impacted by noise [6]. Hence, we summarize the effects of noisy instances for classifying user phone call behavior as follows:

- Create unnecessary classification rules that are not interesting to the users and make the rule-set larger.
- The complexity of the classifiers and the number of necessary training samples may increase.
- The presence of noisy training instances is more likely to cause over-fitting for the decision tree classifier and thus decrease it's accuracy.

According to [24], the performance of the classifier depends on two significant factors: (1) the quality of the training data, and (2) the competence of learning algorithm. Therefore, identification and elimination of the noisy instances from a training dataset are required to ensure the quality of the training data before applying learning technique in order to achieve better classification accuracy.

NBC is the most popular technique to detect noisy instances from a training dataset, as it is attributed to the independence assumption and the use of conditional probabilities [2,5]. Farid et al. [5] have proposed a naive Bayes classifier based noise detection technique for multi-class classification tasks. This technique finds the noisy instances from a training dataset using a naive Bayes classifier and removes these instances from the training set before constructing a decision tree learning for making decisions. In their approach, they identify all the misclassified instances from the training dataset using NBC and consider these instances as noise. However, some of these misclassified instances might represent true *behavioral patterns* of individuals. Therefore, such a strong assumption regarding noisy instances more likely to decrease the *classification accuracy* of mining phone call behavior.

In this paper, we address the above mentioned issue for identifying noisy instances and propose an *improved noise detection technique* based on the naive Bayes classifier for effectively classifying mobile users' phone call behaviors. In our approach, we first calculate the conditional probability for all the instances using naive Bayes classifier and Laplace-estimator. After that we dynamically determine a *noise threshold* according to individual's unique behavioral patterns. Finally, the (misclassified) instances that can't satisfy this threshold are selected as noise. As individual's phone call behavioral patterns are not identical in the real life, this threshold for identifying noisy instances changes dynamically according to the behavior of individuals. To measure the effectiveness of our technique for classifying user phone call behavior, we employ a prominent classification algorithm - decision tree. Our approach aims to improve the existing naive Bayes classifier based noise detection technique [5] for classifying phone call behavior of individuals.

The contributions are summarized as follows:

- We determine a *noise threshold* dynamically according to individual's unique behavioral patterns.
- We propose an *improved noise detection technique* based on naive Bayes classifier for effectively classifying mobile users' phone call behaviors.
- Our experiments on real mobile phone datasets show that this technique is more effective than existing technique for classifying user phone call behavior.

The rest of the paper is organized as follows. We review the naive Bayes classifier and Laplacian estimator in Sects. 2 and 3 respectively. We present our approach in Sect. 4. We report the experimental results in Sect. 5. Finally, Sect. 6 concludes this the paper and highlights the future work.

2 Naive Bayes Classifier

A naive Bayes classifier (NBC) is a simple probabilistic based method, which can predict the class membership probabilities [2,9]. It has two main advantages: (a) easy to use, and (b) only one scan of the training data is required for probability generation. A naive Bayes classifier can easily handle missing attribute values by simply omitting the corresponding probabilities for those attributes when calculating the likelihood of membership for each class. It also requires the class conditional independence, i.e., the effect of an attribute on a given class is independent of those of other attributes.

Let D be a training set of data instances and their associated class labels. Each instance is represented by an n-dimensional attribute vector, $X = \{x_1, x_2, \ldots, x_n\}$, depicting n measurements made on the instance from n attributes, respectively, $\{A_1, A_2, \ldots, A_n\}$. Suppose that there are m classes, $\{C_1, C_2, \ldots, C_m\}$. For a test instance, X, the classifier will predict that X belongs to the class with the highest conditional probability, conditioned on X. That is, the naive Bayes classifier predicts that the instance X belongs to the class C_i, if and only if -

$$P(C_i|X) > P(C_j|X) \text{ for } 1 \leq j \leq m, j \neq i$$

The class C_i for which $P(C_i|X)$ is maximized is called the Maximum Posteriori Hypothesis.

$$P(C_i|X) = \frac{P(X|C_i)P(C_i)}{P(X)} \qquad (1)$$

In Bayes theorem shown in Eq. (1), as $P(X)$ is a constant for all classes, only $P(X|C_i)P(C_i)$ needs to be maximized. If the class prior probabilities are not known, then it is commonly assumed that the classes are likely equal, that is, $P(C_1) = P(C_2) = \ldots = P(C_m)$, and therefore we would maximize $P(X|C_i)$. Otherwise, we maximize $P(X|C_i)P(C_i)$. The class prior probabilities are calculated by $P(C_i) = |C_{i,D}|/|D|$, where $|C_{i,D}|$ is the number of training instances of class C_i in D. To compute $P(X|C_i)$ in a dataset with many attributes is extremely computationally expensive. Thus, the naive assumption of class-conditional independence is made in order to reduce computation in evaluating $P(X|C_i)$. This presumes that the attributes' values are conditionally independent of one another, given the class label of the instance, i.e., there are no dependence relationships among attributes. Thus, Eqs. (2) and (3) are used to produce $P(X|C_i)$.

$$P(X|C_i) = \prod_{k=1}^{n} P(x_k|C_i) \qquad (2)$$

$$P(X|C_i) = P(x_1|C_i) \times P(x_2|C_i) \times \ldots \times P(x_n|C_i) \qquad (3)$$

In Eq. (2), x_k refers to the value of attribute A_k for instance X. Therefore, these probabilities $P(x_1|C_i), P(x_2|C_i), \ldots, P(x_n|C_i)$ can be easily estimated from the training instances. If the attribute value, A_k, is categorical, then $P(x_k|C_i)$ is the number of instances in the class $C_i \in D$ with the value x_k for A_k, divided by $|C_{i,D}|$, i.e., the number of instances belonging to the class $C_i \in D$.

To predict the class label of instance X, $P(X|C_i)P(C_i)$ is evaluated for each class $C_i \in D$. The naive Bayes classifier predicts that the class label of instance X is the class C_i, if and only if -

$$P(X|C_i)P(C_i) > P(X|C_j)P(C_j) \text{ for } 1 \leq j \leq m \text{ and } j \neq i$$

In other words, the predicted class label is the class C_i for which $P(X|C_i)P(C_i)$ is the maximum.

3   Laplacian Estimation

As in naive Bayes classifier, we calculate $P(X|C_i)$ as the product of the probabilities $P(x_1|C_i) \times P(x_2|C_i) \times \ldots \times P(x_n|C_i)$, based on the independence assumption and class conditional probabilities, we will end up with a probability value of zero for some $P(x|C_i)$ if attribute value x is never observed in the training data for class C_i. Therefore, Eq. (3) becomes zeros for such attribute value regardless the values of other attributes. Thus, naive Bayes classifier cannot predict the class of such test instance. Laplace estimate [1] is usually employed to scale up the values by smoothing factor. In Laplace-estimate, the class probability is defined as:

$$P(C = c_i) = \frac{n_c + k}{N + n \times k} \qquad (4)$$

where n_c is the number of instances satisfying $C = c_i$, N is the number of training instances, n is the number of classes and $k = 1$.

Let's consider a phone call behavior example, for the behavior class 'reject' in the training data containing 1000 instances, we have 0 instance with *relationship* = *unknown*, 990 instances with *relationship* = *friend*, and 10 instances with *relationship* = *mother*. The probabilities of these contexts are 0, 0.990 (from 990/1000), and 0.010 (from 10/1000), respectively. On the other hand, according to Eq. (4), the probabilities of these contexts would be as follows:

$$\frac{1}{1003} = 0.001, \frac{991}{1003} = 0.988, \frac{11}{1003} = 0.011$$

In this way, we obtain the above non-zero probabilities (rounded up to three decimal places) respectively using Laplacian-estimation. The "new" probability estimates are close to their "previous" counterparts, and these values can be used for further processing.

4   Noise Detection Technique

In this section, we discuss our noise detection technique in order to effectively classify user phone call behavior. Figure 1 shows the block diagram of our noise detection technique.

In order to detect noise, we use naive Bayes classifier (NBC) [9] as the basis for noise identification. Using NBC, we first calculate the conditional probability for each attribute by scanning the training data. Table 1 shows an example of the mobile phone dataset. Each instance contains four attribute values (e.g., time, location, situation, and relationship between caller and callee) and corresponding phone call behavior. Tables 2 and 3 report the prior probabilities for each behavior class and conditional probabilities for each attribute value, respectively for this dataset. Using these probabilities, we calculate the conditional probability for each instance. As NBC was implemented under the independence assumption, it estimates zero probabilities if the conditional probability for a single attribute is zero. In such cases, we use Laplace-estimator [1] to estimate the conditional probability of any of the attribute value.

Once we have calculated conditional probability for each instance, we differentiate between the purely classified instances and misclassified instances using these values. "Purely classified" instances are those for which the predicted class

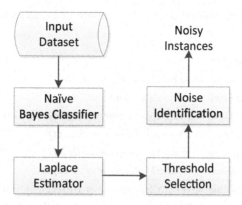

Fig. 1. A block diagram of noise detection technique

Table 1. Sample mobile phone dataset

Day[Time-Segment]	Location	Situation	Relationship	User behavior
Fri[S1]	Office	Meeting	Friend	Reject
Fri[S1]	Office	Meeting	Colleague	Reject
Fri[S1]	Office	Meeting	Boss	Accept
Fri[S1]	Office	Meeting	Friend	Reject
Fri[S2]	Home	Dinner	Friend	Accept
Wed[S1]	Office	Seminar	Unknown	Reject
Wed[S1]	Office	Seminar	Colleague	Reject
Wed[S1]	Office	Seminar	Mother	Accept
Wed[S2]	Home	Dinner	Unknown	Accept

Table 2. Prior probabilities for each behavior class generated using the mobile phone dataset

Probability	Value
P(behavior = Reject)	5/9
P(behavior = Accept)	4/9

and the original class is same. If different class found then these are "misclassified" instances. After that, we generate the instances groups by taking into account all the distinct probabilities as separate group values. Figure 2 shows an example of instances groups $G1, G2, G3$ for the instances $X_1, X_2, \ldots, X_{10}$ where $G1$ consists of 5 instances with probability $p1$, $G2$ consists of 3 instances with probability $p2$ and finally $G3$ consists of 3 instances with probability $p3$. We then identify the group among the purely classified instances for which the probability is minimum. This minimum probability is considered as "noise-threshold".

Table 3. Conditional probabilities for each attribute value calculated using the mobile phone dataset

Probability	Value
$P(DayTime = Fri[S1]|behavior = Reject)$	3/5
$P(DayTime = Fri[S1]|behavior = Accept)$	1/4
$P(DayTime = Fri[S2]|behavior = Reject)$	0/5
$P(DayTime = Fri[S2]|behavior = Accept)$	1/4
$P(DayTime = Wed[S1]|behavior = Reject)$	2/5
$P(DayTime = Wed[S1]|behavior = Accept)$	1/4
$P(DayTime = Wed[S2]|behavior = Reject)$	0/5
$P(DayTime = Wed[S2]|behavior = Accept)$	1/4
$P(Location = Office|behavior = Reject)$	5/5
$P(Location = Office|behavior = Accept)$	2/4
$P(Location = Home|behavior = Reject)$	0/5
$P(Location = Home|behavior = Accept)$	2/4
$P(Situation = Meeting|behavior = Reject)$	3/5
$P(Situation = Meeting|behavior = Accept)$	1/4
$P(Situation = Seminar|behavior = Reject)$	2/5
$P(Situation = Seminar|behavior = Accept)$	1/4
$P(Situation = Dinner|behavior = Reject)$	0/5
$P(Situation = Dinner|behavior = Accept)$	2/4
$P(Relationship = Friend|behavior = Reject)$	2/5
$P(Relationship = Friend|behavior = Accept)$	1/4
$P(Relationship = Colleague|behavior = Reject)$	2/5
$P(Relationship = Colleague|behavior = Accept)$	0/4
$P(Relationship = Boss|behavior = Reject)$	0/5
$P(Relationship = Boss|behavior = Accept)$	1/4
$P(Relationship = Mother|behavior = Reject)$	0/5
$P(Relationship = Mother|behavior = Accept)$	1/4
$P(Relationship = Unknown|behavior = Reject)$	1/5
$P(Relationship = Unknown|behavior = Accept)$	1/4

Finally, the instances in misclassified list, for those probabilities are less than the noise threshold, are identified as noise.

The process for identifying noise is set out in Algorithm 1. Input data includes training dataset: $D = X_1, X_2, \ldots, X_n$, which contains a set of training instances and their associated class labels and output data is the list of noisy instances. For each class, we calculate the prior probabilities $P(C_i)$ (line 2). After that for each attribute value, we calculate the class conditional probabilities $P(A_i|C_i)$

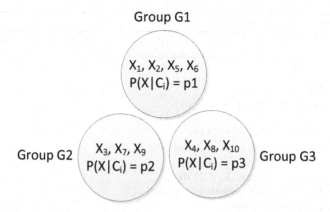

Fig. 2. An example of instances-group based on probability

(line 5). For each training instance, we calculate the conditional probabilities $P(X_i|C_i)$ (line 8). We then check whether it is non-zero. If we get zero probabilities, we then recalculate the conditional probabilities $P(X_i|C_i)$ using Laplacian Estimator (line 11). Based on these probability values, we then check whether the instances are misclassified or purely classified and store all misclassified instances $misClass_{list}$ (line 14) with corresponding probabilities in $misPro_{list}$ (line 15). Similarly, we also store all purely classified instances $pureClass_{list}$ (line 18) with corresponding probabilities in $purePro_{list}$ (line 19). We then identify the minimum probability from $purePro_{list}$ as noise threshold (line 22). As we aim to identify the noise list we check the conditional probabilities $(X_i|C_i)$ in $misPro_{list}$ for all instances. If any instance fails to satisfy this threshold then we store that instance as noise and store into $noise_{list}$ (line 26). Finally this algorithm returns a set of noisy instances $noise_{list}$ (line 29) for a particular dataset.

Rather than arbitrarily determine the threshold, our algorithm dynamically identifies the noise threshold according to individual's behavioral patterns and identify noisy instances based on this threshold. As individual's phone call behavioral patterns are not identical in the real life this noise-threshold for identifying noisy instances changes dynamically according to individual's unique behavioral patterns.

5   Experiments

In this section, we describe our experimental setup and the phone log datasets used in experiment. We also present an experimental evaluation comparing our proposed noise detection technique and the existing naive Bayes classifier based noise detection technique [5] for classifying user phone call behavior.

Algorithm 1. Noise Detection

Data: Training dataset: $D = X_1, X_2, ..., X_n$ // Training dataset, D, which contains a set of training instances and their associated class labels.

Result: noise list: $noise_{list}$

1 **foreach** *class, $C_i \in D$* **do**
2 | Find the prior probabilities, $P(C_i)$.
3 **end**
4 **foreach** *attribute value, $A_i \in D$* **do**
5 | Find the class conditional probabilities, $P(A_i|C_i)$.
6 **end**
7 **foreach** *training instance, $X_i \in D$* **do**
8 | Find the conditional probability, $P(X_i|C_i)$
9 | **if** $P(X_i|C_i) == 0$ **then**
10 | | //use Laplacian Estimator
11 | | recalculate the conditional probability, $P(X_i|C_i)$ using Laplacian Estimator
12 | **end**
13 | **if** X_i *is misclassified* **then**
14 | | $misClass_{list} \leftarrow X_i$
15 | | $misPro_{list} \leftarrow P(X_i|C_i)$ // store the probabilities for all misclassified instances.
16 | **end**
17 | **else**
18 | | $pureClass_{list} \leftarrow X_i$
19 | | $purePro_{list} \leftarrow P(X_i|C_i)$ // store the probabilities for all purely classified instances.
20 | **end**
21 **end**
22 $T_{noise} = findMIN(purePro_{list})$ // use as noise threshold
23 **foreach** *instance, $x_i \in misClass_{list}$* **do**
24 | Find the conditional probability, $P(X_i|C_i)$ from $misPro_{list}$
25 | **if** $P(X_i|C_i) < T_{noise}$ **then**
26 | | $noise_{list} \leftarrow X_i$ // store instances as noise.
27 | **end**
28 **end**
29 return $noise_{list}$

5.1   Experimental Setup

We have implemented our noise detection technique (Algorithm 1) and existing naive Bayes classifier based technique [5] in Java programming language and executed them on a Windows PC with an Intel Core I5 CPU (3.20 GHz) and 8 GB memory. In order to measure the classification accuracy, we first eliminate the noisy instances identified by noise identification technique from the training dataset, and then apply the decision-tree classifier [15] on the noise-free dataset. The reason for choosing the decision tree as a classifier is that decision tree is the most popular classification algorithm in data mining [21,22]. The code for the basic versions of the decision tree classifier is adopted from Weka, which is an open source data mining software [7].

5.2 Dataset

We have conducted experiments on phone log datasets of five individual mobile phone users (randomly selected from Massachusetts Institute of Technology (MIT) Reality Mining dataset [4]). We extract 7-tuple information of the call record for each phone user from the datasets: date of call, time of call, call-type, call duration, location, relationship, call ID. These datasets contain three types of phone call behavior, e.g., incoming, missed and outgoing. As can be seen, the user's behavior in accepting and rejecting calls are not directly distinguishable in incoming calls in the dataset. As such, we derive accept and reject calls by using the call duration. If the call duration is greater than 0 then the call has been accepted; if it is equal to 0 then the call has been rejected [16]. We also pre-process the temporal data in mobile phone log as it is continuous and numeric. For this, we use BOTS technique [16] for producing behavior-oriented time segments. Table 4 describes each dataset of the individual mobile phone user.

Table 4. Datasets descriptions

Dataset	Contexts	Instances	Behavior classes
User 04	Temporal, location, relationship	5119	Accept, reject, missed, outgoing
User 23	Temporal, location, relationship	1229	Accept, reject, missed, outgoing
User 26	Temporal, location, relationship	3255	Sccept, reject, missed, outgoing
User 33	Temporal, location, relationship	635	Accept, reject, missed, outgoing
User 51	Temporal, location, relationship	2096	Accept, reject, missed, outgoing

5.3 Evaluation Metric

In order to measure the classification accuracy, we compare the classified response with the actual response (i.e., the ground truth) and compute the accuracy in terms of:

– Precision: ratio between the number of phone call behaviors that are correctly classified and the total number of behaviors that are classified (both correctly and incorrectly). If TP and FP denote true positives and false positives then the formal definition of precision is:

$$Precision = \frac{TP}{TP + FP} \qquad (5)$$

– Recall: ratio between the number of phone call behaviors that are correctly classified and the total number of behaviors that are relevant. If TP and FN denote true positives and false negatives then the formal definition of recall is:

$$Recall = \frac{TP}{TP + FN} \tag{6}$$

– F-measure: a measure that combines precision and recall is the harmonic mean of precision and recall. The formal definition of F-measure is:

$$Fmeasure = 2 * \frac{Precision * Recall}{Precision + Recall} \tag{7}$$

5.4   Evaluation Results

To evaluate our approach, we employ the 10-fold cross validation on each dataset. In k fold cross-validation, the initial data are randomly partitioned into k mutually exclusive subsets or "folds", $d_1, d_2, \ldots, d_k$, each of which has an approximately equal size. Training and testing are performed k times. In iteration i, the partition d_i is reserved as the test set, and the remaining partitions are collectively used to train the classifier. Therefore, the 10-fold cross validation breaks data into 10 sets of size N/10. It trains the classifier on 9 sets and tests it using the remaining one set. This repeats 10 times and we take a mean accuracy rate. For classification, the accuracy estimate is the total number of correct classifications from the k-iterations, divided by the total number of instances in the initial dataset. To show the effectiveness of our technique, we compare the accuracy of both the existing naive Bayes classifier based noise detection approach (NBC) [5] and our proposed dynamic threshold based approach, in terms of precision, recall and f-measure.

Table 5. The accuracies of existing naive Bayes classifier based approach (NBC)

Dataset	Precision	Recall	F-measure
User 04	0.91	0.30	0.45
User 23	0.83	0.84	0.83
User 26	0.89	0.51	0.65
User 33	0.80	0.85	0.80
User 51	0.78	0.78	0.78

Tables 5 and 6 show the experimental results for five individual mobile phone users' datasets using the existing naive Bayes classifier based noise detection approach and our dynamic threshold based approach respectively. From Tables 5 and 6, we find that our approach consistently outperforms previous NBC-based approach for all individuals in terms of precision, recall and F-measure. In addition to compare individual level, we also show the relative comparison of average precision, average recall and average F-measure for all the five different datasets in Fig. 3.

Table 6. The accuracies of our proposed dynamic threshold based approach

Dataset	Precision	Recall	F-measure
User 04	0.89	0.70	0.78
User 23	0.84	0.84	0.84
User 26	0.92	0.72	0.80
User 33	0.82	0.86	0.81
User 51	0.86	0.85	0.85

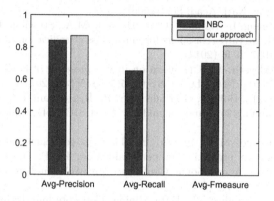

Fig. 3. Effectiveness comparison results

The experimental results for a collection of users show that our approach consistently outperforms the NBC-based approach. The reason is that instead of treating all misclassified instances as noise we identify true noisy instances from misclassified list using a noise threshold. We determine this noise threshold for each individual dataset as it varies according to individual's unique behavioral patterns. As a result, our technique improves the classification accuracy while classifying phone call behavior of individual mobile phone users.

6 Conclusion and Future Work

In this paper, we have presented an approach to detecting and eliminating noisy instances from mobile phone data in order to improve the classification accuracy. Our approach dynamically determines the noise threshold according to individual's behavioral patterns. For this, we employ both the naive Bayes classifier and Laplacian estimator. Experimental results on multi-contextual phone call log datasets indicate that compare to the NBC-based approach, our approach improves the classification accuracy in terms of precision, recall and F-measure.

In future work, we plan to investigate the effect of noise on confidence threshold to produce association rules. We will extend our noise detection technique to produce confidence-based association rules of individual mobile phone users in multi-dimensional contexts.

References

1. Cestnik, B., et al.: Estimating probabilities: a crucial task in machine learning. In: ECAI, vol. 90, pp. 147–149 (1990)
2. Chen, J., Huang, H., Tian, S., Qu, Y.: Feature selection for text classification with Naïve Bayes. Expert Syst. Appl. **36**(3), 5432–5435 (2009)
3. Daza, L., Acuna, E.: An algorithm for detecting noise on supervised classification. In: Proceedings of WCECS 2007, the 1st World Conference on Engineering and Computer Science, pp. 701–706 (2007)
4. Eagle, N., Pentland, A., Lazer, D.: Infering social network structure using mobile phone data. Proc. Natl. Acad. Sci. (2006)
5. Farid, D.M., Zhang, L., Rahman, C.M., Hossain, M.A., Strachan, R.: Hybrid decision tree and Naïve Bayes classifiers for multi-class classification tasks. Expert Syst. Appl. **41**(4), 1937–1946 (2014)
6. Frénay, B., Verleysen, M.: Classification in the presence of label noise: a survey. IEEE Trans. Neural Netw. Learn. Syst. **25**(5), 845–869 (2014)
7. Hall, M., Frank, E., Holmes, G., Pfahringer, B., Reutemann, P., Witten, I.H.: The WEKA data mining software: an update. ACM SIGKDD Explor. Newsl. **11**(1), 10–18 (2009)
8. Halvey, M., Keane, M.T., Smyth, B.: Time based segmentation of log data for user navigation prediction in personalization. In: Proceedings of the 2005 IEEE/WIC/ACM International Conference on Web Intelligence, pp. 636–640. IEEE Computer Society (2005)
9. Han, J., Pei, J., Kamber, M.: Data Mining: Concepts and Techniques. Elsevier, Amsterdam (2011)
10. John, G.H., Langley, P.: Estimating continuous distributions in Bayesian classifiers. In: Proceedings of the Eleventh Conference on Uncertainty in Artificial Intelligence, pp. 338–345. Morgan Kaufmann Publishers Inc. (1995)
11. Mehrotra, A., Hendley, R., Musolesi, M.: PrefMiner: mining user's preferences for intelligent mobile notification management. In: Proceedings of the 2016 ACM International Joint Conference on Pervasive and Ubiquitous Computing, pp. 1223–1234. ACM (2016)
12. Ozer, M., Keles, I., Toroslu, H., Karagoz, P., Davulcu, H.: Predicting the location and time of mobile phone users by using sequential pattern mining techniques. Comput. J. **59**(6), 908–922 (2016)
13. Pejovic, V., Musolesi, M.: InterruptMe: designing intelligent prompting mechanisms for pervasive applications. In: Proceedings of the 2014 ACM International Joint Conference on Pervasive and Ubiquitous Computing, pp. 897–908. ACM (2014)
14. Phithakkitnukoon, S., Dantu, R., Claxton, R., Eagle, N.: Behavior-based adaptive call predictor. ACM Trans. Auton. Adapt. Syst. (TAAS) **6**(3), 21 (2011)
15. Ross Quinlan, J.: C4.5: programs for machine learning. Mach. Learn. (1993)
16. Sarker, I.H., Colman, A., Kabir, M.A., Han, J.: Behavior-oriented time segmentation for mining individualized rules of mobile phone users. In: 2016 IEEE International Conference on Data Science and Advanced Analytics (DSAA), Montreal, Canada, pp. 488–497. IEEE (2016)
17. Sarker, I.H., Kabir, M.A., Colman, A., Han, J.: An effective call prediction model based on noisy mobile phone data. In: Proceedings of the 2017 ACM International Joint Conference on Pervasive and Ubiquitous Computing: Adjunct. ACM (2017)

18. Srinivasan, V., Moghaddam, S., Mukherji, A.: MobileMiner: mining your frequent patterns on your phone. In: ACM International Joint Conference on Pervasive and Ubiquitous Computing. ACM (2014)
19. Witten, I.H., Frank, E.: Data Mining: Practical Machine Learning Tools and Techniques. Morgan Kaufmann, Burlington (2005)
20. Witten, I.H., Frank, E., Trigg, L.E., Hall, M.A., Holmes, G., Cunningham, S.J.: Weka: practical machine learning tools and techniques with Java implementations (1999)
21. Wu, C.-C., Chen, Y.-L., Liu, Y.-H., Yang, X.-Y.: Decision tree induction with a constrained number of leaf nodes. Appl. Intell. **45**(3), 673–685 (2016)
22. Wu, X., Kumar, V., Quinlan, J.R., Ghosh, J., Yang, Q., Motoda, H., McLachlan, G.J., Ng, A., Liu, B., Philip, S.Y., et al.: Top 10 algorithms in data mining. Knowl. Inf. Syst. **14**(1), 1–37 (2008)
23. Zhu, H., Chen, E.: Mining mobile user preferences for personalized context-aware recommendation. ACM Trans. Intell. Syst. Technol. **5**(4) (2014)
24. Zhu, X., Wu, X.: Class noise vs. attribute noise: a quantitative study. Artif. Intell. Rev. **22**(3), 177–210 (2004)

Big Data

Big Data

A Two-Sample Kolmogorov-Smirnov-Like Test for Big Data

Hien D. Nguyen[✉]

Department of Mathematics and Statistics, La Trobe University,
Bundoora, VIC 3086, Australia
h.nguyen5@latrobe.edu.au

Abstract. Exploratory data analysis (EDA) is an important component of modern data analysis and data mining. The Big Data setting has made many traditional and useful EDA tools impractical and ineffective. Among such useful tools is the two-sample Kolmogorov-Smirnov (TS-KS) goodness-of-fit (GoF) test for assessing whether or not two samples arose from the same population. A TS-KS like testing procedure is constructed using chunked and averaged (CA) estimation paradigm. The procedure is named the TS-CAKS GoF test. Distributed and streamed implementations of the TS-CAKS procedure are discussed. The consistency of the TS-CAKS test is proved. A numerical study is provided to demonstrate the effectiveness and computational efficiency of the procedure.

Keywords: Big Data · Chunked-and-average estimator
Hypothesis testing · Kolmogorov-Smirnov test

1 Introduction

Exploratory data analysis (EDA), termed by Tukey (1962), is the discipline of data science that is concerned with the summarization and characterization of data sets, rather than the use of data for drawing formal inferences regarding predefined hypotheses; see Tukey (1977) for a book-length treatment of the topics and techniques that make up EDA. In the modern setting, EDA techniques are core disciplinary tools for data analysts and data scientists; see Dasu and Johnson (2003) and Myatt and Johnson (2014) for modern perspectives on EDA and its interplay with data mining.

The contemporary Big Data setting has introduced numerous issues that prevent the effective deployment of traditional EDA tools. These issues include the distribution of data, the arrival of data in streams, rather than the availability of data in batch, and the shear scale and volume of available data for analysis. Within this context, Buoncristiano et al. (2015) describes the availability of Big Data-suitable FDA techniques as being an important direction of research in data mining. This sentiment is echoed in Mecca (2016) where the availability of fast operators for computing statistical properties is seen as a fundamental requirement for modern database exploration. A particular point is made by

© Springer Nature Singapore Pte Ltd. 2018
Y. L. Boo et al. (Eds.): AusDM 2017, CCIS 845, pp. 89–106, 2018.
https://doi.org/10.1007/978-981-13-0292-3_6

Buoncristiano et al. (2015) regarding the need for fast and efficient sample testing and comparison algorithms.

Let $\mathbf{X} = \{X_j\}_{j=1}^N$ and $\mathbf{Y} = \{Y_j\}_{j=1}^N$ be a pair of IID samples of $N \in \mathbb{N}$ observations. Suppose that each observation X_j and Y_j arise from distributions with continuous cumulative distribution functions (CDFs) F and G, respectively, for each

$$j \in [N] = \{1, 2, \ldots, N\}.$$

We denote realizations from the samples as $\mathbf{x} = \{x_j\}_{j=1}^N$ and $\mathbf{y} = \{y_j\}_{j=1}^N$. Given $\mathbf{x}$ and $\mathbf{y}$, among the most common tasks that one may wish to conduct, in order to characterize the data, is to inspect whether or not $F = G$. The overarching class of tools for conduction such assessments are known as two-sample goodness-of-fit (TS-GoF) tests.

The most popular TS-GoF test, in practice, is that of Smirnov (1939), and generally referred to as the two-sample Kolmogorov-Smirnov (TS-KS) test. The TS-KS test assess the validity of the null hypothesis $H_0 : F = G$ against the alternative $H_1 : F \neq G$. Upon observation of $\mathbf{x}$ and $\mathbf{y}$, one conducts the TS-KS test by computing the KS test statistic

$$K(\mathbf{x}, \mathbf{y}) = \sup_{t \in \mathbb{D}} \left| \hat{F}(t; \mathbf{x}) - \hat{G}(t; \mathbf{y}) \right|, \tag{1}$$

where

$$\hat{F}(t; \mathbf{x}) = N^{-1} \sum_{j=1}^N \mathbb{I}(x_j \leq t)$$

and

$$\hat{G}(t; \mathbf{y}) = N^{-1} \sum_{j=1}^N \mathbb{I}(y_j \leq t)$$

are the empirical distribution functions of $\mathbf{x}$ and $\mathbf{y}$, respectively. Here, $\mathbb{D}$ is the common domain of F and G, and $\mathbb{I}(A)$ is the indicator variable, which takes value 1 if A is true and 0 otherwise.

When N is small (e.g. $N \leq 100$, as is implemented in the ks.test() function in the R programming environment (R Core Team 2016), the probability of observing a value equal or great than (1) (i.e. the p-value) can be computed via the method of Kim and Jennrich (1973). If N is large, then we must use the asymptotic approximation:

$$\lim_{N \to \infty} \mathbb{P}\left[(N/2)^{1/2} K_N(\mathbf{x}, \mathbf{y}) \leq t | H_0 \right] \tag{2}$$

$$= 1 - 2 \sum_{k=1}^{\infty} (-1)^{k+1} \exp\left(-2k^2 t^2\right),$$

instead (cf. DasGupta 2008, Theorem 26.6). The problem with the TS-KS test— in the Big Data context—is that it must be applied in batch (when conducted

naively), it cannot be conducted in parallel, and it is too computationally intensive for application to large data sets (e.g. $N \geq 10^9$), even when using the
asymptotic approximation (2).

There has been some recent research into Big Data-appropriate implementations of the TS-KS test. For example, dos Reis et al. (2016) explored the use
of incremental TS-KS tests for identifying drifts between two populations over
time. Their algorithm is made stream-suitable by the clever use of a tree structure to either insert or delete observations from the currently computed TS-KS
statistic. Another development is due to Lall (2015). Here, quantile sketches are
used to approximate the TS-KS statistic, using only a fraction of total number of
pairs N, in a stream. The algorithm thus addresses much of the issues regarding
the application of the TS-KS test in the Big Data context. Unfortunately, due
to the deterministic approximation nature of the test statistic computation, the
decision rules are made conservative (i.e. the rejection region of the test is set
to guarantee tests of size at least $\alpha \in (0, 1)$, but may potentially be greater).
Streamed KS-like testing procedures for change detection were also considered
in Kifer et al. (2004).

A caveat to both of the aforementioned articles is the lack of application to
very large data sets. That is, for example, the methods of Lall (2015) and Reis et
al. (2016) are only assessed on data of sizes $N = 10^5$ and $N = 10^3$, respectively.
Thus, we find that there is a gap in the literature concerning the development
of GoF tests in such contexts, as well as the context of distributed TS-KS tests,
which is also not addressed in either of referenced articles.

Recently, Nguyen (2017b) has constructed a one-sample KS-like test for
assessing deviations of some data $\mathbf{x}$ from an ideal hypothesized CDF F. The KS-
like test of Nguyen (2017b) was constructed via a chunked-and-averaged (CA)
estimation paradigm of Li et al. (2013) and Matloff (2016), and is thus termed
CAKS. The CAKS testing procedure was developed for streamed data, but can
also be used for GoF testing of large data sets on one or multiple distributed
processing units. The procedure was also theoretically proved to be consistent
in the sense of having the correct size when the null is true and rejecting any
quantifiable deviation away from the null, when it is false, as N gets large.

In this paper, we follow the work of Nguyen (2017b) and construct a two-
sample CAKS (TS-CAKS) procedure for testing H_0 versus H_1. We demonstrate
that our test shares the consistency property of the CAKS procedure and discuss
how it can be applied to distributed and streamed data. A numerical study is also
conducted in order to demonstrate the computational efficiency and effectiveness
of the TS-CAKS procedure.

The paper proceeds as follows. The CA paradigm for distributed and
streamed estimation is introduced in Sect. 2. The TS-CAKS test is derived in
Sect. 3. A consistency result is presented and proved in Sect. 4. Numerical simulations are presented in Sect. 5. Finally, conclusions are drawn in Sect. 6.

2   CA Estimators

Let

$$\theta = \theta\left(F, G\right)$$

be some parameter of interest, which is dependent on the CDFs F and G. Given that we receive samples $\mathbf{x}$ and $\mathbf{y}$ in batch, suppose that we can compute the batch estimate

$$\hat{\theta}_N = \hat{\theta}\left(\mathbf{x}, \mathbf{y}\right),$$

where $\hat{\theta}$ is some appropriate algorithm.

Let z be a placeholder for either x or y. Now, consider that we arrange the elements of $\mathbf{x}$ and $\mathbf{y}$ as

$$
\begin{aligned}
& z_{11}, z_{12}, \ldots, z_{1C}, \\
& z_{21}, z_{22}, \ldots, z_{2C}, \\
& \ldots, \ldots, \ldots, \ldots, \\
& z_{R1}, z_{R2}, \ldots, z_{RC},
\end{aligned}
\tag{3}
$$

where $z_{rc} = z_{(r-1)C+c}$, for each $r \in [R]$ and $c \in [C]$, and $N = R \times C$. We say that $\mathbf{x}_c = \{x_{rc}\}_{r=1}^R$ and $\mathbf{y}_c = \{y_{rc}\}_{r=1}^R$ are the cth chunks of their respective samples. We further write the cth chunked estimator of θ as

$$\hat{\theta}_{R,c} = \hat{\theta}\left(\mathbf{x}_c, \mathbf{y}_c\right).$$

Using the definitions that are given, we can write the CA estimator of θ as

$$\bar{\theta}_{R,C} = C^{-1}\sum_{c=1}^{C}\hat{\theta}_{R,c}. \tag{4}$$

Let

$$\hat{\Theta}_{R,c} = \hat{\Theta}\left(\mathbf{X}_c, \mathbf{Y}_c\right)$$

and

$$\bar{\Theta}_{R,C} = C^{-1}\sum_{c=1}^{C}\hat{\Theta}_{R,c}$$

be random versions of $\hat{\theta}_{R,c}$ and $\bar{\theta}_{R,C}$, respectively. Assume that the mean

$$\mu_R = \mathbb{E}\left(\hat{\Theta}_{R,c}\right)$$

and variance

$$\sigma_R^2 = \mathbb{E}\left(\hat{\Theta}_{R,c}^2\right) - \mu_R^2$$

exist and denote convergence in distribution by $\leadsto$. We say that ζ is a standard normal random variable if its probability generating process is normal with mean zero and variance one. The following result is available from Li et al. (2013).

Lemma 1. *Assume that* **X** *and* **Y** *are IID samples from some probability models with CDFs F and G, respectively. If* μ_R *and* σ_R^2 *exist, then*

$$\frac{\bar{\Theta}_{R,C} - \mu_R}{\sigma_R/C^{1/2}} \rightsquigarrow \zeta,$$

as $C \to \infty$, *where* ζ *is a standard normal random variable.*

Remark 1. Note that for fixed R, $\bar{\Theta}_{R,C}$ is a consistent estimator of θ if and only if $\mu_R = \theta$ (i.e. $\hat{\Theta}_{R,c}$ is unbiased, for each c). This is not true in general and is not true in our application of Lemma 1. Fortunately, we do not require $\bar{\Theta}_{R,C}$ to be consistent and only require that it converges to a finite limit. This point is elaborated upon in the sequel.

Remark 2. Apart from Lemma 1, there are numerous results regarding the properties of the CA estimator under various conditions of growth in R and N, as well as dependence structure in **X** and **Y**. We direct the interest reader to Li et al. (2013) and Matloff (2016) regarding IID data, and Nguyen & McLachlan (2017) regarding dependent data.

3   The TS-CAKS Test

Following the notation of the previous section, we define θ for the TS-KS test as

$$\theta (F, G) = \sup_{t \in \mathbb{D}} |F(t) - G(t)|$$

and estimate it via the batch statistic

$$\hat{\theta} (\mathbf{x}, \mathbf{y}) = K (\mathbf{x}, \mathbf{y}).$$

Given arrangement (3), we can also define the cth chunked estimator as

$$\hat{\theta}_{R,c} (\mathbf{x}_c, \mathbf{y}_c) = K (\mathbf{x}_c, \mathbf{y}_c)$$

and thus also the CA estimator $\bar{\theta}_{R,C}$. As before, let $\hat{\Theta}_{R,c}$ and $\bar{\Theta}_{R,C}$ be random versions of $\hat{\theta}_{R,c}$ and $\bar{\theta}_{R,C}$, respectively. We notice that $\hat{\Theta}_{R,c}$ is a measurable function and its CDF, for any R, can be directly computed directly using the approach of Kim & Jennrich (1973). Thus $\hat{\Theta}_{R,c}$ has all of its moments and therefore μ_R and σ_R^2 exist. Lemma 1 provides the following result.

Theorem 1. *Let*

$$\bar{\Theta}_{R,C} = C^{-1} \sum_{c=1}^{C} K (\mathbf{X}_c, \mathbf{Y}_c)$$

and assume that **X** *and* **Y** *are IID samples from some probability models with CDF F = G. If*

$$\mu_R = \mathbb{E} [K (\mathbf{X}_c, \mathbf{X}_c)]$$

and

$$\sigma_R^2 = \mathbb{E}\left[K^2\left(\mathbf{X}_c, \mathbf{Y}_c\right)\right] - \mu_R^2,$$

then

$$\frac{\bar{\Theta}_{R,C} - \mu_R}{\sigma_R / C^{1/2}} \rightsquigarrow \zeta,$$

as $C \to \infty$, where ζ is a standard normal random variable.

We have demonstrated that $\bar{\Theta}_{R,C}$ is asymptotically normal under H_0 and thus we can construct a Z-test for deviations away from the null. Unfortunately, as noted in Remark 1, μ_R cannot simply be replaced by any parameter value as it is inconsistent. Similarly, we do not know the value of σ_R^2. Fortunately, we can perform a large-scale Monte Carlo simulation to produce well-estimated values for μ_R and σ_R^2 under H_0 at useful values of R. The values reported in Table 1 are computed using 10^5 random samples each. The samples $\mathbf{X}$ and $\mathbf{Y}$ are drawn from uniform distributions on the unit interval. Due to the invariance of $K\left(\mathbf{X}, \mathbf{Y}\right)$ under the null, this is purely a practical choice (cf. DasGupta 2008, Sect. 26.9). We report the values of the means and variances of the normalized statistic $(R/2)^{1/2}\bar{\Theta}_{R,C}$ in order to conform with the probability statement from (2).

Table 1. Values for $(R/2)^{1/2}\mu_R$ and $(R/2)\sigma_R^2$, estimated via 10^5 Monte Carlo replicates.

R	$(R/2)^{1/2}\mu_R$	$(R/2)\sigma_R^2$
5.00E+01	0.821104	0.067535
1.00E+02	0.834514	0.067634
2.00E+02	0.844277	0.067798
5.00E+02	0.853303	0.067814
1.00E+03	0.857670	0.067744
2.00E+03	0.860759	0.067705
5.00E+03	0.863662	0.067785
1.00E+04	0.865045	0.067764
2.00E+04	0.865953	0.067724
5.00E+04	0.867568	0.067639
1.00E+05	0.867774	0.067960
2.00E+05	0.867781	0.067760
5.00E+05	0.868099	0.067669
1.00E+06	0.868321	0.067831
$R \to \infty$	0.868731	0.067773

Remark 3. In cases where R is larger than the values that are available on the table, (2) can be used to obtain an asymptotic mean and variance via the formulas

$$\mathbb{E}\left[(N/2)^{1/2} K\left(\mathbf{X}, \mathbf{Y}\right)\right] \approx (\pi/2)^{1/2} \log 2$$

and

$$\mathbb{E}\left(\left[(N/2)^{1/2} K\left(\mathbf{X}, \mathbf{Y}\right)\right]^2\right) \approx \pi^2/2;$$

see Wang et al. (2003) for details.

Let $\alpha \in (0, 1)$ be a predetermined size (or Type I error) of the test what we wish to conduct. Further, let ζ_α be the $(1 - \alpha)$th quantile of the standard normal distribution. Using the asymptotic normality of $\bar{\Theta}_{R,C}$ from Theorem 1, we can reject H_0, in favor of H_1 if

$$\bar{\theta}_{R,C} > \mu_R + \zeta_\alpha \frac{\sigma_R}{\sqrt{C}}. \tag{5}$$

Remark 4. As claimed in the introduction, the test of form (5) is suitable for both distributed and streamed data. In the case where data is distributed, one can consider a scenario whereupon the N pairs of observations are distributed across $B \leq C$ processing units (PUs). In such a case, we can consider each of C chunks of R observations as being distributed among the B PUs and each of the chunked estimates $\hat{\theta}_{R,c}$ can be computed independently and in parallel. The CA estimator can then be estimated by a master PU once each of the C chunked estimates have been computed on their respective PUs. Thus, unlike the TS-KS test, the data can be distributed among a number of locations and only the chunked estimators need to be communicated to a master PU.

In the case where the data are streamed, we can consider that the total stream has N observations and at each time period c, we receive R observations from the stream. For streamed data, Nguyen (2017a) suggest an iterative equation for (4) of form

$$\bar{\theta}_{R,c} = \frac{\bar{\theta}_{R,c-1} + \hat{\theta}_{R,c}}{c}, \tag{6}$$

for $c \in [C]$, where $\bar{\theta}_{R,0} \in \mathbb{R}$ is arbitrarily. At each time period c, we observe that Eq. (6) requires only the R pairs of observations from $\mathbf{x}_c$ and $\mathbf{y}_c$, and the previous iterate $\bar{\theta}_{R,c-1}$. Thus, the online stream implementation of the TS-CAKS test has space complexity of $O(R)$, where R is a fixed constant that does not dependent on N. We believe this to be a remarkable result since the algorithm utilizes all N pairs of observations that are available.

Remark 5. We note that a normal approximation for the TS-KS test had also been considered by Kim (1969). However, unlike our exploitation of the asymptotic normality of the average of finitely sampled KS statistics, Kim (1969) instead utilized normal approximations to match moments with finite sampling distributions in order to obtain accuracy gains on the asymptotic approximation (2).

4   Consistency

For some fixed size α, let $\{T_N\}_{N=1}^{\infty}$ be a sequence of test statistics and let $\{\chi_N\}_{N=1}^{\infty}$ be a sequence of critical values. Suppose that the test statistic is a function of $\mathbf{X}$ and $\mathbf{Y}$, where X_j and Y_j are randomly drawn from distributions F and G, respectively. For each $N \in \mathbb{N}$, conduct a hypothesis test of $H_0 : (F, G) \in \Omega_0$ versus $H_1 : (F, G) \in \Omega_1$ by rejecting H_0 in favor of H_1 if $T_N > \chi_N$. Adapting the notation of DasGupta (2008, Definition 24.1), we say that the sequence of test statistics $\{T_N\}_{N=1}^{\infty}$ and critical values $\{\chi_N\}_{N=1}^{\infty}$ are consistent for H_0 versus H_1 if

$$\mathbb{P}\left(T_N > \chi_N\right) \to \alpha$$

as $N \to \infty$, for all $(F, G) \in \Omega_0$, and

$$\mathbb{P}\left(T_N > \chi_N\right) \to 1$$

as $N \to \infty$, for all $(F, G) \in \Omega_1$.

Consistency therefore implies that the test has the correct size, as more observations are obtained, and that the test will reject the null under any alternative in Ω_1, given enough data. We adapt the following consistency result from DasGupta (2008 Theorem 24.2).

Lemma 2. *For some fixed size $\alpha \in (0, 1)$, let $\{T_N\}_{N=1}^{\infty}$ be a sequence of test statistics and let $\{\chi_N\}_{N=1}^{\infty}$ be a sequence of critical value for the testing of $H_0 : (F, G) \in \Omega_0$ versus $H_1 : (F, G) \in \Omega_1$, such that*

$$\mathbb{P}\left(T_N > \chi_N\right) \to \alpha$$

as $N \to \infty$, for all $(F, G) \in \Omega_0$. Assume that (i) T_N converges in probability to some functional $\mu(F, G)$ for any $(F, G) \in \Omega_0 \cup \Omega_1$; (ii)

$$\mu(F, G) = \mu_0$$

for all $(F, G) \in \Omega_0$, and for all $(F, G) \in \Omega_1$,

$$\mu(F, G) > \mu_0;$$

and (iii) T_N is asymptotically normal with mean μ_0 and variance σ_0^2/N, where $0 < \sigma_0^2 < \infty$. If the test rejects H_0 in favor of H_1 when $T_N > \chi_N$, for each N, then the sequence of tests is consistent against alternatives of form H_1.

In order to establish consistency, we must validate the assumptions of Lemma 2 in the context of the TS-CAKS test. First, we acknowledge that there is a separate TS-CAKS test for each fixed R. Next, in the TS-CAKS context,

$$\Omega_0 = \{(F, G) : F = G\}$$

and

$$\Omega_1 = \{(F, G) : F \neq G, \mathbb{E}\left[K\left(\mathbf{X}_c, \mathbf{Y}_c\right)\right] > \mu_R\}.$$

For any fixed R, we have a sequence of TS-CAKS test statistics $\{\bar{\Theta}_{R,C}\}_{C=1}^{\infty}$, which is compared against the sequence of critical values $\{\chi_{R,C}\}_{C=1}^{N}$, where

$$\chi_{R,C} = \mu_R + \zeta_\alpha C^{1/2} \sigma_R.$$

The asymptotic normality of $\bar{\Theta}_{R,C}$ from Theorem 1 along with the invariance of the KS test statistic immediately validates the criterion that

$$\mathbb{P}\left(\bar{\Theta}_{R,C} > \chi_{R,C}\right) \to \alpha$$

as $N \to \infty$ under H_0. Next, to validate (i), we simply note, as before, that $K(\mathbf{X}_c, \mathbf{Y}_c)$ is bounded to the unit interval, thus a mean always exist and is a function of the population distributions F and G. To validate (ii), we note that under the invariance principle, the mean of the test statistic is

$$\mu(F, G) = \mu_0$$

under H_0. Although it is intuitive that the condition that $\mu(F, G) > \mu_0$ should be met under any interesting alternative, where $F \neq G$, we have not found a satisfactory proof of this result. Thus we must take this assumption as given. Finally, (iii) is validated again by taking $\mu_0 = \mu_J$, $\sigma_0^2 = \sigma_J^2$ and replacing N by C. Here, σ_J^2 exists since $K(\mathbf{X}_c, \mathbf{Y}_c)$ is bounded. Thus, we can state the following theorem regarding the TS-CAKS test.

Theorem 2. *Assume that* $\mathbf{X}$ *and* $\mathbf{Y}$ *are IID samples from distributions with CDFs* F *and* G, *respectively. For any fixed* $R \in \mathbb{N}$ *and size* $\alpha \in (0, 1)$, *the sequence of TS-CAKS rules of form (5), for* $C \in \mathbb{N}$, *is consistent for testing* $H_0 : F = G$ *against* $H_1 : F \neq G$ *and*

$$\mathbb{E}\left[K(\mathbf{X}_c, \mathbf{Y}_c)\right] > \mu_R.$$

5   Numerical Study

We conduct a numerical study using a set of three simulation scenarios S1–S3. All computations conducted for the study are performed on a MacBook Pro with a 2.2 GHz Intel Core i7 processor, 16 GB of 1600 MHz DDR3 memory, and a 500 GB SSD. Furthermore, implementations of all computational processes and algorithms are performed via the R programming language and interpreter environment (R Core Team Core Team 2016).

In all three simulation scenarios, S1–S3, the same basic setup is used. That is,

$$C \in \{100, 1000\}$$

chunks of observations of sizes

$$R \in \bigcup_{i=2}^{5} \{1, 2, 5\} \times 10^i \cup \{50, 10^6\}$$

are simulated from distributions that depend on the simulation scenarios. The CAKS test, using the $N = R \times C$, is conducted for the null hypothesis $H_0 : F = G$. In all three scenarios, we simulate $\mathbf{x}$ from a standard normal distribution. In scenario S1, we simulate $\mathbf{y}$ from a normal distribution with variance 1 but varying means

$$\mu \in \{0, 0.001, 0.01\}.$$

In scenario S2, we simulate $\mathbf{y}$ from a normal distribution with mean 0 but with varying variances

$$\sigma^2 \in \{0.8^2, 0.9^2, 0.95^2\}.$$

In scenario S3, we simulate $\mathbf{y}$ from a t distribution with degrees of freedom

$$\mathrm{df} \in \{10, 50, 100\}.$$

For each of the scenarios and at each of the variable values, we replicate the CAKS test 100 times. From the replicates, we note the total number of rejections at the size $\alpha = 0.1$ as well as the average time taken to conduct the tests. The results of the numerical study are reported in Tables 2–7. The average times are given in seconds and timing was conducted using the proc.time() function.

5.1   Results

Upon inspection of Tables 2–7, we observe the that the TS-CAKS test behaves as per the conclusion of Theorem 2. That is, we observe that under H_0 (i.e. the first column of Tables 2 and 5), the number of rejections tended to be within an interval around 10. Noting that the 95% margin of error for Wald confidence interval for a sample of 100 with population proportion $\alpha = 0.1$ is ±6, we can conclude that all of the values are in a reasonable bound and that each test has the correct size.

Next in the alternative columns of Table 2 and 5, and all columns of Tables 3, 4, 6, and 7, we observe that the numbers of rejections increase with C. This provides empirical affirmation of the asymptotic power approaching one as the number of observations increases, as concluded in Theorem 2. The increasing in power appears to be exhibited as a function of both increases in R or C. As with all hypothesis testing procedures, we also observe that the further an alternative hypothesis is from the null, the easier it is to distinguish.

The timing columns from Tables 2–7 appear to indicate that the computational times required for the TS-CAKS test increases approximately linearly on average with the number of chunks C. This is an excellent outcome as we can be confident that the procedure will scale proportionally with data size. Considering that each of the timing measurements actually correspond to the collective computational time required over $C = 1000$ chunks, we can further infer that the computational demand of the TS-CAKS procedure is actually much lighter when considering its parallel or streamed implementations.

Finally, using some further simulations, we establish that the average computational times required by the ks.test() function for 1.00E+04, 1.00E+05,

Table 2. Results from simulation scenario S1 when $C = 100$. Reject and Time indicate number of test rejections and average computational time in seconds, respectively.

$\mu = 0$			$\mu = 0.001$		
R	Reject	Time	R	Reject	Time
5.00E+01	9	8.80E−03	5.00E+01	9	7.10E−03
1.00E+02	5	1.14E−02	1.00E+02	5	9.54E−03
2.00E+02	9	1.71E−02	2.00E+02	4	1.45E−02
5.00E+02	10	3.41E−02	5.00E+02	8	2.94E−02
1.00E+03	12	6.38E−02	1.00E+03	11	5.46E−02
2.00E+03	9	1.30E−01	2.00E+03	13	1.09E−01
5.00E+03	12	3.46E−01	5.00E+03	12	2.76E−01
1.00E+04	11	7.96E−01	1.00E+04	7	5.75E−01
2.00E+04	10	1.73E+00	2.00E+04	9	1.26E+00
5.00E+04	12	4.80E+00	5.00E+04	10	3.74E+00
1.00E+05	12	9.14E+00	1.00E+05	13	7.84E+00
2.00E+05	9	2.11E+01	2.00E+05	20	1.69E+01
5.00E+05	16	7.32E+01	5.00E+05	49	5.11E+01
1.00E+06	9	1.48E+02	1.00E+06	86	1.22E+02

$\mu = 0.01$		
R	Reject	Time
5.00E+01	15	7.12E−03
1.00E+02	5	9.53E−03
2.00E+02	9	1.47E−02
5.00E+02	13	3.03E−02
1.00E+03	16	5.48E−02
2.00E+03	22	1.07E−01
5.00E+03	40	2.79E−01
1.00E+04	86	5.74E−01
2.00E+04	99	1.26E+00
5.00E+04	100	3.74E+00
1.00E+05	100	7.85E+00
2.00E+05	100	1.74E+01
5.00E+05	100	5.54E+01
1.00E+06	100	1.40E+02

1.00E+06, 1.00E+07, and 1.00E+08 number of observations are 1.29E−02, 1.20E−01, 1.62E+00, 3.64E+01, and 6.16E+02, respectively (for standard normal data, averaged over 100 times). We can compare these results directly with the results from the first column of Tables 2 and 5. We notice that TS-CAKS is consistently an order of magnitude faster than the ks.test() function in R, for

Table 3. Results from simulation scenario S2 when $C = 100$. Reject and Time indicate number of test rejections and average computational time in seconds, respectively.

$\sigma^2 = 0.8^2$			$\sigma^2 = 0.9^2$		
R	Reject	Time	R	Reject	Time
5.00E+01	95	8.40E−03	5.00E+01	37	8.03E−03
1.00E+02	100	1.12E−02	1.00E+02	53	1.05E−02
2.00E+02	100	1.68E−02	2.00E+02	87	1.57E−02
5.00E+02	100	3.33E−02	5.00E+02	100	3.10E−02
1.00E+03	100	6.21E−02	1.00E+03	100	5.75E−02
2.00E+03	100	1.26E−01	2.00E+03	100	1.13E−01
5.00E+03	100	3.34E−01	5.00E+03	100	2.90E−01
1.00E+04	100	7.73E−01	1.00E+04	100	6.01E−01
2.00E+04	100	1.67E+00	2.00E+04	100	1.38E+00
5.00E+04	100	4.65E+00	5.00E+04	100	4.24E+00
1.00E+05	100	8.81E+00	1.00E+05	100	8.29E+00
2.00E+05	100	1.91E+01	2.00E+05	100	2.02E+01
5.00E+05	100	6.65E+01	5.00E+05	100	6.76E+01
1.00E+06	100	1.64E+02	1.00E+06	100	1.66E+02

$\sigma^2 = 0.95^2$		
R	Reject	Time
5.00E+01	15	8.13E−03
1.00E+02	20	1.06E−02
2.00E+02	22	1.56E−02
5.00E+02	65	3.08E−02
1.00E+03	94	5.76E−02
2.00E+03	100	1.13E−01
5.00E+03	100	2.91E−01
1.00E+04	100	6.05E−01
2.00E+04	100	1.39E+00
5.00E+04	100	4.32E+00
1.00E+05	100	8.84E+00
2.00E+05	100	2.06E+01
5.00E+05	100	6.55E+01
1.00E+06	100	1.68E+02

all N except for the 1.00E+04 case. Furthermore, it must be mentioned that we could not compute a single instance of the 1.00E+09 observations case using ks.test() due to memory limitations. Thus, it appears that TS-CAKS is both faster than the usual batch KS test and also scales better with increases in the number of observations.

Table 4. Results from simulation scenario S3 when $C = 100$. Reject and Time indicate number of test rejections and average computational time in seconds, respectively.

df = 10			df = 50		
R	Reject	Time	R	Reject	Time
5.00E+01	14	7.96E−03	5.00E+01	5	8.02E−03
1.00E+02	22	1.09E−02	1.00E+02	9	1.10E−02
2.00E+02	19	1.70E−02	2.00E+02	8	1.71E−02
5.00E+02	51	3.51E−02	5.00E+02	12	3.50E−02
1.00E+03	95	6.58E−02	1.00E+03	10	6.56E−02
2.00E+03	100	1.29E−01	2.00E+03	10	1.29E−01
5.00E+03	100	3.31E−01	5.00E+03	20	3.29E−01
1.00E+04	100	6.80E−01	1.00E+04	46	6.78E−01
2.00E+04	100	1.52E+00	2.00E+04	80	1.50E+00
5.00E+04	100	4.33E+00	5.00E+04	100	4.32E+00
1.00E+05	100	9.02E+00	1.00E+05	100	9.00E+00
2.00E+05	100	1.92E+01	2.00E+05	100	1.92E+01
5.00E+05	100	5.72E+01	5.00E+05	100	5.70E+01
1.00E+06	100	1.37E+02	1.00E+06	100	1.80E+02

df = 100		
R	Reject	Time
5.00E+01	8	8.75E−03
1.00E+02	8	1.17E−02
2.00E+02	10	1.80E−02
5.00E+02	9	3.67E−02
1.00E+03	8	6.83E−02
2.00E+03	8	1.36E−01
5.00E+03	15	3.49E−01
1.00E+04	12	7.15E−01
2.00E+04	24	1.70E+00
5.00E+04	53	5.23E+00
1.00E+05	92	1.01E+01
2.00E+05	100	2.39E+01
5.00E+05	100	7.96E+01
1.00E+06	100	1.85E+02

Table 5. Results from simulation scenario S1 when $C = 1000$. Reject and Time indicate number of test rejections and average computational time in seconds, respectively.

$\mu = 0$			$\mu = 0.001$		
R	Reject	Time	R	Reject	Time
5.00E+01	11	6.42E−02	5.00E+01	7	7.00E−02
1.00E+02	10	8.73E−02	1.00E+02	8	8.77E−02
2.00E+02	12	1.36E−01	2.00E+02	6	1.36E−01
5.00E+02	3	2.86E−01	5.00E+02	11	2.84E−01
1.00E+03	10	5.48E−01	1.00E+03	7	5.42E−01
2.00E+03	4	1.10E+00	2.00E+03	12	1.08E+00
5.00E+03	7	2.89E+00	5.00E+03	13	2.80E+00
1.00E+04	5	6.34E+00	1.00E+04	11	5.93E+00
2.00E+04	9	1.39E+01	2.00E+04	12	1.37E+01
5.00E+04	10	3.90E+01	5.00E+04	18	3.88E+01
1.00E+05	9	7.87E+01	1.00E+05	35	7.92E+01
2.00E+05	10	1.81E+02	2.00E+05	50	1.79E+02
5.00E+05	10	5.46E+02	5.00E+05	87	5.46E+02
1.00E+06	4	1.72E+03	1.00E+06	100	1.71E+03
$\mu = 0.01$					
R	Reject	Time			
5.00E+01	11	6.91E−02			
1.00E+02	12	8.95E−02			
2.00E+02	8	1.35E−01			
5.00E+02	16	2.85E−01			
1.00E+03	28	5.43E−01			
2.00E+03	58	1.08E+00			
5.00E+03	97	2.81E+00			
1.00E+04	100	5.93E+00			
2.00E+04	100	1.38E+01			
5.00E+04	100	3.85E+01			
1.00E+05	100	7.92E+01			
2.00E+05	100	1.83E+02			
5.00E+05	100	5.45E+02			
1.00E+06	100	1.71E+03			

Table 6. Results from simulation scenario S2 when $C = 1000$. Reject and Time indicate number of test rejections and average computational time in seconds, respectively.

$\sigma^2 = 0.8^2$			$\sigma^2 = 0.9^2$		
R	Reject	Time	R	Reject	Time
5.00E+01	100	8.36E−02	5.00E+01	73	7.44E−02
1.00E+02	100	1.13E−01	1.00E+02	100	9.75E−02
2.00E+02	100	1.73E−01	2.00E+02	100	1.46E−01
5.00E+02	100	3.45E−01	5.00E+02	100	3.00E−01
1.00E+03	100	6.51E−01	1.00E+03	100	5.73E−01
2.00E+03	100	1.32E+00	2.00E+03	100	1.15E+00
5.00E+03	100	3.52E+00	5.00E+03	100	2.97E+00
1.00E+04	100	8.18E+00	1.00E+04	100	6.40E+00
2.00E+04	100	1.77E+01	2.00E+04	100	1.52E+01
5.00E+04	100	4.70E+01	5.00E+04	100	4.19E+01
1.00E+05	100	8.89E+01	1.00E+05	100	8.27E+01
2.00E+05	100	1.75E+02	2.00E+05	100	1.60E+02
5.00E+05	100	5.28E+02	5.00E+05	100	5.82E+02
1.00E+06	100	1.46E+03	1.00E+06	100	2.03E+03

$\sigma^2 = 0.95^2$		
R	Reject	Time
5.00E+01	18	7.40E−02
1.00E+02	30	9.55E−02
2.00E+02	73	1.43E−01
5.00E+02	100	2.97E−01
1.00E+03	100	5.63E−01
2.00E+03	100	1.12E+00
5.00E+03	100	2.93E+00
1.00E+04	100	6.22E+00
2.00E+04	100	1.44E+01
5.00E+04	100	4.05E+01
1.00E+05	100	8.01E+01
2.00E+05	100	1.64E+02
5.00E+05	100	5.03E+02
1.00E+06	100	1.77E+03

Table 7. Results from simulation scenario S3 when $C = 1000$. Reject and Time indicate number of test rejections and average computational time in seconds, respectively.

df = 10			df = 50		
R	Reject	Time	R	Reject	Time
5.00E+01	19	7.54E−02	5.00E+01	9	8.23E−02
1.00E+02	33	9.96E−02	1.00E+02	10	1.10E−01
2.00E+02	66	1.56E−01	2.00E+02	12	1.68E−01
5.00E+02	100	3.32E−01	5.00E+02	11	3.54E−01
1.00E+03	100	6.34E−01	1.00E+03	18	6.74E−01
2.00E+03	100	1.26E+00	2.00E+03	28	1.34E+00
5.00E+03	100	3.26E+00	5.00E+03	64	3.45E+00
1.00E+04	100	6.85E+00	1.00E+04	98	7.27E+00
2.00E+04	100	1.56E+01	2.00E+04	100	1.65E+01
5.00E+04	100	4.32E+01	5.00E+04	100	4.58E+01
1.00E+05	100	8.94E+01	1.00E+05	100	9.43E+01
2.00E+05	100	1.95E+02	2.00E+05	100	2.22E+02
5.00E+05	100	7.67E+02	5.00E+05	100	8.17E+02
1.00E+06	100	1.54E+03	1.00E+06	100	1.62E+03

df = 100		
R	Reject	Time
5.00E+01	10	8.19E−02
1.00E+02	8	1.10E−01
2.00E+02	5	1.68E−01
5.00E+02	11	3.52E−01
1.00E+03	12	6.72E−01
2.00E+03	22	1.33E+00
5.00E+03	24	3.44E+00
1.00E+04	37	7.25E+00
2.00E+04	67	1.65E+01
5.00E+04	100	4.56E+01
1.00E+05	100	9.42E+01
2.00E+05	100	1.89E+02
5.00E+05	100	8.60E+02
1.00E+06	100	1.54E+03

6   Conclusions

EDA is an important component of modern data analysis and data mining. Unfortunately, the Big Data setting has introduced numerous problems that has made many traditional EDA tools inapplicable, inefficient, or ineffective. Among these tools is the TS-KS GoF test for assessing whether or not two samples arose from the same population.

We introduced an alternative to the TS-KS test using a CA estimator construction. We named our procedure the TS-CAKS GoF test and discussed how it could be implemented in both the distributed and streamed data settings. Furthermore, we prove that our test is consistent in the large-sample setting, when the alternative hypothesis is identifiable from the null.

A numerical study was conducted using three simulation scenarios. Across the three scenarios, we demonstrated that the TS-CAKS procedure maintains the correct size under the null hypothesis and that—as concluded by the theoretical results—the power of the test increases to one as the number of observations increases. Lastly, we show that the test appears to demonstrate linear scaling in computational time, on average, and that it is approximately an order of magnitude faster than the default batch TS-KS test implementation in the R programming language.

A minor dissatisfaction that a reader might have regarding the TS-CAKS test is that unlike the TS-KS test, it requires that the samples $\mathbf{x}$ and $\mathbf{y}$ be of equal size N. We do not anticipate that this will be a major issue in the Big Data setting as one always tend to have enough data from two populations in order to organize into equal sized samples. Similarly, in a streaming scenario, unless the rates of data flow are drastically different, one can always wait for the arrival of enough data in order to form two equal samples. The only practical situation when one may encounter a need for an unbalanced TS-KS test—in the Big Data context—is when comparing between a common and rare population. In such cases, there is a high likelihood that the populations are different and thus the TS-CAKS test may be unnecessary. When it is necessary, one can always fall back on the traditional TS-KS implementation or that of Lall (2015).

Acknowledgements. The author is personally supported by Australian Research Council grant number DE170101134.

References

Buoncristiano, M., Mecca, G., Quitarelli, E., Roveri, M., Santoro, D., Tanca, L.: Database challenges for exploratory computing. ACM SIGMOD Rec. **44**, 17–22 (2015)

DasGupta, A.: Asymptotic Theory of Statistics and Probability. Springer, New York (2008). https://doi.org/10.1007/978-0-387-75971-5

Dasu, T., Johnson, T.: Exploratory Data Mining and Data Cleaning. Wiley, New York (2003)

dos Reis, D., Flach, P., Matwin, S., Batista, G.: Fast unsupervised online drift detection using incremental Kolmogorov-Smirnov test. In: ACM SIGKDD International Conference on Knowledge Disocvery and Data Mining XXII. ACM (2016)

Kifer, D., Ben-David, S., Gehrke, J.: Detecting change in data streams. In: Proceedings of the 30th VLDB Conference (2004)

Kim, P.J.: On the exact and approximate sampling distribution of the two sample Kolmogorov-Smirnov criterion D_{mn}, $m \leq n$. J. Am. Stat. Assoc. **64**, 1625–1637 (1969)

Kim, P.J., Jennrich, R.I.: Selected tables in mathematical statistics 1, chapter tables of the exact sampling distribution of the two-sample Kolmogorov-Smirnov criterion D_{mn}, $m \leq n$, pp. 80–129. Institute of Mathematical Statistics (1973)

Lall, A.: Data streaming algorithm for the Kolmogorov-Smirnov test. In: Proceedings of the IEEE International Conference on Big Data, pp. 95–104 (2015)

Li, R., Lin, D.K.J., Li, B.: Statistical inference in massive data sets. Appl. Stoch. Models Bus. Ind. **29**, 399–409 (2013)

Matloff, N.: Software alchemy: turning complex statistical computations into embarrassingly-parallel ones. J. Stat. Softw. **71**, 1–15 (2016)

Mecca, G.: Database exploration: problems and opportunities. In: IEEE 32rd International Conference on Data Engineering Workshop, pp. 153–156 (2016)

Myatt, G.J., Johnson, W.P.: Making Sense of Data I: A Practical Guide to Exploratory Data Analysis and Data Mining. Wiley, New York (2014)

Nguyen, H.D.: A simple online parameter estimation technique with asymptotic guarantees. arXiv:1703.07039 (2017a)

Nguyen, H.D.: A stream-suitable Kolmogorov-Smirnov-type test for Big Data analysis. arXiv:1704.03721 (2017b)

Nguyen, H.D., McLachlan, G.J.: Chunked-and-averaged estimators for vector parameters. arXiv:1612.06492 (2017)

R Core Team: R: a language and environment for statistical computing. R Foundation for Statistical Computing (2016)

Smirnov, N.V.: Estimating the deviation between the empirical distribution functions of two independent samples. Bulletin de l'Universite de Moscou, **2**, 3–16 (1939)

Tukey, J.W.: The future of data analysis. Ann. Math. Stat. **33**, 1–67 (1962)

Tukey, J.W.: Exploratory Data Analysis. Addison-Wesley, Reading (1977)

Wang, J., Tsang, W.W., Marsaglia, G.: Evaluating Kolmogorov's distribution. J. Stat. Softw. **8**, 18 (2003)

Exploiting Redundancy, Recurrency and Parallelism: How to Link Millions of Addresses with Ten Lines of Code in Ten Minutes

Yuhang Zhang[✉], Tania Churchill, and Kee Siong Ng

Australian Transaction Reports and Analysis Centre, Canberra, Australia
yuhang.zhang@austrac.gov.au

Abstract. Accurate and efficient record linkage is an open challenge of particular relevance to Australian Government Agencies, who recognise that so-called wicked social problems are best tackled by forming partnerships founded on large-scale data fusion. Names and addresses are the most common attributes on which data from different government agencies can be linked. In this paper, we focus on the problem of address linking. Linkage is particularly problematic when the data has significant quality issues. The most common approach for dealing with quality issues is to standardise raw data prior to linking. If a mistake is made in standardisation, however, it is usually impossible to recover from it to perform linkage correctly. This paper proposes a novel algorithm for address linking that is particularly practical for linking large disparate sets of addresses, being highly scalable, robust to data quality issues and simple to implement. It obviates the need for labour intensive and problematic address standardisation. Empirical results show that approximately 91% of the generated links created by matching two large address datasets from two government agencies, were correct. Finally, we demonstrate that the linking can be performed in under 10 min, with 10 lines of code.

Keywords: Record linkage · Address linking

1 Introduction

Efficient record linkage is an important step in large-scale automated data fusion. Data fusion is a problem of increasing significance in the context of Australia's whole-of-government approach to tackling our most pressing social issues - including terrorism and welfare fraud - by combining and analysing datasets from multiple government agencies. Outside of personal identifiers like tax-file numbers and driver's licenses, names and addresses are the two most important attributes on which disparate datasets are matched. Whereas the problem of linking names is well-studied and there are specialised similarity

© Springer Nature Singapore Pte Ltd. 2018
Y. L. Boo et al. (Eds.): AusDM 2017, CCIS 845, pp. 107–122, 2018.
https://doi.org/10.1007/978-981-13-0292-3_7

measures like Jaro-Winkler for names [7], not a great deal is known in the literature [4] about best practices for matching addresses, especially address data with significant quality issues. Listed here are some address-specific challenges for efficient record linkage:

- **Incompleteness:** incomplete addresses that have no street type, no suburb name, no postcode, etc are common in address data.
- **Inconsistent Formats:** the structure of addresses can be different between countries, regions and languages. People can use variants to denote the same address, e.g., unit 1 of 2 Elizabeth Street, 1/2 Elizabeth St, and U-1 2 Elisabeth Str all denote the same address. The presence of foreign characters in international addresses can also introduce issues.
- **Errors:** Wrong street types, invalid postcodes, non-matching suburb-postcode pairs, and various misspellings are widely seen in address data.
- **Unsegmented Addresses:** Depending on the source, addresses can be captured as a single line of text with no explicit structural information.

These data quality issues may make two equivalent addresses look different and, by chance, make two different addresses look similar.

The most common way to tackle data quality issues is to standardise raw addresses before the linking operation [4]. Address standardisation usually includes two types of operations: parsing and transforming. With the parsing operation, addresses are parsed into semantic components, such as street, suburb, state, and country. For example, if an address contains three numbers they are in order unit number, street number, and postcode. In the transforming operation, variants of the same entity are transformed to a canonical format and typos are removed, e.g., transforming Street, St, and Str all to Street.

The issue with standardisation is that it is in itself a challenging problem. For example, *Service Centre St George* might be interpreted as a business name *Service Centre of Saint George*; or a street name and a suburb name *Service Centre Street, George*; or a different business name *Service Centre of Street George*; or a suburb name and a state name *Service Centre, Saint George*. Three numbers in an address can also be street number, level number in a high-rise, and postcode. Interpreting an address is by nature ambiguous.

Address standardisation can be done using a rule-based system, or it can be done using machine learning approaches like Hidden Markov Models [5,6]. Ongoing research is still being undertaken to improve standardisation accuracy [9]. Perhaps the biggest drawback of address standardisation is that if a mistake is made during standardisation, it is usually hard to recover from it to perform linkage correctly. Rule-based standardisation also tends to be specific to the individual dataset, failing to generalise well.

Using Redundancy to Avoid Standardisation

Instead of standardising raw addresses into canonical forms, we rely on the redundancy in addresses to resolve data quality issues.

We say an address contains redundancy if an incomplete representation is sufficient to uniquely identify this address. For example, if there is only one building in Elizabeth St that has Unit 123, then *U123 45 Elizabeth St* as an address contains redundancy, because specifying street number 45 is not really necessary. Redundancy exists widely in addresses. Not every suburb is covered by postcode 2600. Not every state has a street named Elizabeth. As an extreme example, three numbers like 18 19 5600, might be enough to identify a unique address globally, as long as no other addresses contain these three numbers simultaneously. Note that in this case, we do not even need to know whether 18 is a unit number or a street number.

Our working hypothesis is that address data, in general, contains enough redundancy such that:

1. Each address is still unique even when meta-data distinguishing address components such as street, suburb, and state are missing.
2. Equivalent addresses are still more similar to each other than to irrelevant addresses in the presence of errors or variants.

Our assumptions - which stem from earlier experiments using compressed sensing techniques [2] to represent and link addresses - are really stating that despite the data quality issues in addresses, two addresses, in their raw form, can still be separated/linked if they are different/equivalent. In particular, address segmentation - a problem that is arguably as difficult as the general address-linking problem - and address standardisation are not strictly necessary.

Using Recurrency for Data-Driven Blocking

When linking two large databases, algorithm efficiency is as important as algorithm accuracy. An algorithm that takes days to finish is not only too expensive to deploy, but is also infeasible to repetitively evaluate during development.

Blocking is a widely used technique to improve linkage efficiency. Naïvely, linking two databases containing m and n addresses respectively requires $O(mn)$ comparisons. Most of these comparisons lead to non-matches. To reject these non-matches with a lower cost, one may first partition the raw addresses according to criteria selected by a user. These criteria are called blocking keys, which may be postcode, suburb name, *etc.* During linkage, comparison is only carried out between addresses that fall into the same partitions, based on the assumption that addresses which don't share a blocking key are not a match.

Blocking key selection largely determines the efficiency and completeness of address linkage. If the keys are not meaningful, they will not help find matches and may even slow down the matching process. If too few keys are used, efficiencies won't be gained. If too many keys are used, one may fail to discover all possible links. If different blocking keys do not distribute evenly among the addresses, the largest few partitions will form the bottleneck of linkage efficiency. Moreover, the performance of blocking keys in previous work also depends on the accuracy of address standardisation.

In the spirit of [10], we propose in this paper a data-driven approach to select blocking keys based on their recurrency. These data-driven blocking keys are by design adapted to the database at hand, statistically meaningful as address differentiators, evenly distributed, and provide comprehensive cover to all addresses. Since we implement no standardisation, our blocking keys do not depend on the success of standardisation either.

Implementation on Parallel Platforms

Massively parallel processing databases like Teradata and Greenplum have long supported parallelised SQL that scales to large datasets. Recent advances in large-scale in-database analytics platforms [11,14] have shown us how sophisticated machine learning algorithms can be implemented on top of a declarative language like SQL or MapReduce to scale to petabyte-sized datasets on cluster computing. Building on the same general principle, we propose in this paper a modified inverted index data structure for address linking that can be implemented in less than ten SQL statements and which enjoys tremendous scalability and code maintainability.

Paper Contributions

The paper's contribution is a novel address-linkage algorithm that:

1. links addresses as free-text (including international addresses), obviating the need for labour-intensive and sometimes problematic address standardisation;
2. uses data-driven blocking keys to minimise unnecessary pairwise comparisons, in a way that obviates the need for address segmentation and avoids the usual worst-case scenarios encountered by using a fixed blocking key like suburb or postcode;
3. introduces an extension of the inverted index data structure that allows two large address datasets to be linked efficiently;
4. is practical because of its simplicity, allowing the whole algorithm to be written in less than 10 standard SQL statements; and
5. is scalable when the SQL statements are implemented on top of parallel platforms like the Greenplum Database (open-source parallel PostgreSQL) and Spark.

The algorithm is particularly suitable for integrating large sets of disparate address datasets with minimal manual human intervention. It is also possible to combine the algorithm with a rule-based system to produce a model-averaging system that is more robust than each system in isolation.

The remaining sections of this paper are organised as follows. We first explain how we link a single address to an address database utilising redundancy. We then show how the same algorithm can be carried out in batch taking advantage of recurring address components. We then demonstrate the performance of our algorithm with two address linkage applications, followed by our conclusion.

2 Address as Bag of Tokens

Without subfield structures, an address becomes a bag (or a multiset) of unordered tokens. For example,

No.	Street	Suburb	State	Postcode
513	Elizabeth St	Melbourne	VIC	3000

becomes

$$\{3000, 513, \text{Elizabeth}, \text{Melbourne}, \text{Street}, \text{VIC}\},$$

In this example, we implicitly define a token to be a word, or a maximal character sequence that contains only letters and numerics. We can also define a token to be a single character,

$$\{0, 0, 0, 1, 3, 3, 5, a, b, b, c, e, e, e, e, h, i, l, l, n, o, r, r, s, t, t, t, u, v, z\},$$

a two-word phrase,

$$\{513 \text{ Elizabeth}, \text{ Elizabeth St}, \text{ St Melbourne}, \text{ Melbourne VIC}, \text{ VIC } 3000\}.$$

or generally anything we like. Note that in the above example, two-word phrases preserve pairwise order information in the original address. We can also use two word tokens that do not contain pairwise order information.

Different types of tokens have different distinctiveness powers and different tolerances against data quality issues. To see the difference, note the word token, 'Melbourne', can match to any appearance of 'Melbourne' in other addresses, such as Melbourne Avenue, Mount Melbourne, Melbourne in Canada, *etc.* By contrast, the phrase token, 'Melbourne VIC', can only match the co-occurrence of 'Melbourne' and 'VIC'. The advantage of being distinctive is that we can reduce false matches. The disadvantage, however, is that we may miss a true match if the other address did not include the state information of 'VIC' or included it in a different form, *e.g.*, Victoria.

For the purposes of linkage, we do not need individual tokens to be distinctive. Instead, we want tokens to be tolerant to data quality issues. We lose nothing as long as a bag of tokens as a whole is distinctive enough to identify an address uniquely. However, for matching efficiency we prefer distinctive tokens. We will come back to this topic after we explain how to measure the similarity between two addresses as two bags of tokens.

3 Similarity Between Bags of Tokens

We assess the similarity between two addresses as the similarity between two bags of tokens.

We use Jaccard index to measure the similarity between two bags of tokens. Jaccard index of two sets is defined as the ratio between the number of common elements and the number of total elements.

$$J(T_1, T_2) = \frac{|T_1 \cap T_2|}{|T_1 \cup T_2|} \tag{1}$$

For example, consider two bags of tokens

$$T_1 = \{this, is, an, example\}$$
$$T_2 = \{this, is, another, example\}$$
$$T_1 \cap T_2 = \{this, is, example, this, is, example\}$$
$$T_1 \cup T_2 = \{this, is, an, example, this, is,$$
$$another, example\}$$
$$J(T_1, T_2) = \frac{|T_1 \cap T_2|}{|T_1 \cup T_2|} = \frac{6}{8} = 0.75.$$

As one can see, the Jaccard index between two sets is always in the range between 0 and 1. Here 0 indicates that two sets have nothing in common, and 1 that the two sets are exactly the same. The more common elements two sets share relative to the total number of tokens they have, the larger their Jaccard index is. We say two addresses are equivalent if their Jaccard index exceeds a threshold τ.

We shall see in Sect. 9 that the algorithm admits other similarity functions too.

4   Inverted Index

Naïvely, linking an address to a database requires comparing this particular address against each database address to obtain their similarity. Indexed tokens allow us to do the linking in sublinear time.

We build an inverted index for addresses in the database. An inverted index keeps all the distinct tokens in the database. For each distinct token, the inverted index also keeps references to all the addresses which contain this token.

When a query address arrives, an inverted index allows us to know which database addresses share common tokens with the query address without scanning through the database. More specifically, given a query address, we first break this query address into a bag of tokens Q. If a token is not included in the inverted index, we simply ignore the token. Each remaining token selects a segment from the inverted index. Database addresses appearing on these segments share at least one common token with the query address. We can then count the number of occurrences of each database address C_i on these segments, which gives us the value of $|Q \cap C_i|$ for each i. We then derive the value of $|Q \cup C_i|$ for each i using

$$|Q \cup C_i| = |Q| + |C_i| - |Q \cap C_i|. \tag{2}$$

We can then calculate the Jaccard index between the query address Q and each candidate address C_i using Eq. 1.

With an inverted index, we only compute the Jaccard index between a query address and those database addresses whose Jaccard indexes are non-zero. The efficiency of address linkage therefore depends on the number of addresses that share at least one token with the query address, not the size of the database.

5   Two-Round Linkage

Recall our earlier discussion that tokens of different types have different distinctiveness. The number of database addresses that contain a more distinctive token is by definition smaller than the number of database addresses that contain a less distinctive token. We therefore have better linking efficiency with more distinctive tokens. Yet in return, we may miss more matches due to data quality issues.

To maximise linking efficiency while minimising the number of missed matches, we propose a two-round linkage schema. In the first round, we use distinctive tokens, *e.g.*, phrase tokens, and inverted indexes to shortlist database addresses which have non-zero Jaccard indexes with the query address. In the second round, we compute the Jaccard index between the query and shortlisted addresses using less distinctive tokens to account for data quality issues.

In this way, the distinctive tokens decide which database entries get involved in the linkage. The less distinctive tokens decide the similarity between a query and a database entry. A database entry gets involved as long as it shares a distinctive token with the query. A database entry matches a query if they have enough less-distinctive tokens in common.

The two-round linkage strategy is similar to the one described in [1].

6   A Batch Linkage Algorithm

Quite often, we need to find equivalent addresses between two large databases each containing millions of addresses. Naïvely, we could perform pairwise matching for every combination of addresses. We describe in this section a simple, and possibly novel, extension of the inverted index data structure to allow efficient linking of two large address databases.

To do batch linking between two databases, we build separate inverted indexes for each database. From each inverted index, we eliminate all the tokens that recur more than k times. (More on that soon.) We then join the two inverted indexes by the common tokens they share. Joining a pair of common tokens essentially joins two sets of addresses from two databases, respectively. Every pair of addresses from these two sets is a potential match. Between these pairs, we then compute the Jaccard index to identify true matches.

We eliminate tokens that recur more than k times. If a token is too common, addresses linked by this token are not likely to be a true match. Moreover, examining addresses linked by a common token takes a lot of time, but does not find proportionally more matches. Ignoring these common tokens will not miss many true matches because these matches are usually also linked by some more distinctive tokens.

Linking one address at a time can be seen as a special case of batch linkage, *i.e.*, one of the databases contains only one address. The advantage of batch linkage over performing a single linkage many times is that in batch linkage we join the two inverted indexes only once, instead of many times.

Our batch linkage can be explained in the traditional framework of data linkage, where joining two inverted indexes implements (data-driven) blocking. Nevertheless, there are also some notable differences. Instead of using fixed blocking keys like postcode and suburb, we use tokens as blocking keys. Importantly, deciding which token is used as a blocking key is determined by the data, more specifically its recurring frequency. This allows the algorithm to adapt to characteristics of the specific databases to be matched.

The above extension of inverted indexes applies to the first of the two-round linkage schemes described above. The second round of pairwise Jaccard calculations of shortlisted candidate address pairs is done using the algorithm described in the following section.

Computing Jaccard Index in Linear Time

We first sort the tokens in each set. This can be done efficiently since the number of distinct tokens is small. We then sort the tokens, and read from the two sets at the same time following the rules below:

1. If the two tokens read in are the same, we increase the number of common tokens and the number of total tokens both by 2. We read one more token from each set.
2. If one token is larger than the other, we increase the number of total tokens by 1. We read one more token from the set whose current token is smaller.

We finish reading when either set is exhausted, and add the number of remaining tokens in the other set into the number of total tokens. The division between the number of common tokens and the number of total tokens then provides the Jaccard index.

For small tokens (like characters or 2-g), the time complexity of the algorithm is $O(l + r)$, where l and r denote the number of tokens in the two sets.

SQL Implementation

The full algorithm in (almost ANSI) SQL is listed in Algorithm 1. The SQL code runs on Greenplum and PostgreSQL. The DISTRIBUTED BY keyword in table creation specifies how the rows of a table are stored distributively across a cluster by hashing on the distribution key. The Greenplum database query

Algorithm 1. SQL Code for Batch Linkage

```
1: CREATE TABLE address_db    %% Original address data
   (   address_id bigint,
       address text       )
   DISTRIBUTED BY (address_id);
2: CREATE TABLE address_db_phrase    %% Compute 2-word phrase tokens
   (   address_id bigint,
       token_phrase text   )
   DISTRIBUTED BY (token_phrase);
3: INSERT INTO address_db_phrase
   SELECT address_id,(regexp_matches( regexp_replace(address,'[^A-Z0-9]+',' ','g')
       ,'[A-Z0-9+]+ [A-Z0-9+]+','g'))[1]
   FROM address_db
   UNION
   SELECT address_id,(regexp_matches( regexp_replace( regexp_replace(address,
       '[^A-Z0-9]',' ','g') ,'[A-Z0-9]+','') ,'[A-Z0-9+]+ [A-Z0-9+]+','g'))[1]
   FROM address_db;
4: CREATE TABLE address_db_phrase_inverted    %% Compute inverted index
   (   token_phrase text,
       address_ids bigint[],
       frequency bigint    )
   DISTRIBUTED BY (token_phrase);
5: INSERT INTO address_db_phrase_inverted
   SELECT token_phrase,array_agg(address_id),count(1)
   FROM address_db_phrase
   GROUP BY token_phrase;
6: CREATE TABLE address_db_phrase_matched %% Matched address arrays
   (   token_phrase text,
       address_ids_1 bigint[],
       address_ids_2 bigint[]   )
   DISTRIBUTED BY (token_phrase);
7: %% address_db_phrase_inverted_2 is the second dataset.
   INSERT INTO address_db_phrase_matched
   SELECT l.token_phrase,l.address_ids,r.address_ids
   FROM address_db_phrase_inverted_1 AS l
   INNER JOIN address_db_phrase_inverted_2 AS r
   ON l.token_phrase=r.token_phrase AND l.frequency≤ 100 AND r.frequency≤ 100;
8: CREATE TABLE address_db_proposed_match %% Unnest candidate address pairs
   (   address_id_1 bigint,
       address_id_2 bigint   )
   DISTRIBUTED BY (address_id_1);
9: INSERT INTO address_db_proposed_match
   SELECT DISTINCT address_id_1, unnest(address_ids_2)
   FROM ( SELECT unnest(address_ids_1) AS address_id_1, address_ids_2
               FROM address_db_phrase_matched ) AS tmp;
10: CREATE TABLE address_db_match AS    %% Compute round 2 Jaccard index
   SELECT address_id_1, address_id_2, jaccard(t2.address, t3.address)
   FROM address_db_proposed_match t1,
       address_db_1 t2,
       address_db_2 t3
   WHERE t1.address_id_1 = t2.address_id
   AND t1.address_id_2 = t3.address_id
```

optimiser will exploit the structure of the SQL query and the underlying data distribution to construct optimal execution plans.

With minor modifications, the SQL code can be modified to run on other parallel databases like Teradata and Netezza, and parallel platforms like Spark (using Spark SQL) and Hadoop (using HIVE, HAWQ [3] or Impala [12]). It's also straightforward to implement the algorithm in Scala/Python running natively on Spark.

7   Experiments

We demonstrate the performance of our proposed algorithm in two scenarios: linking an address dataset against a reference address dataset, and linking two arbitrary address datasets. In the first scenario, for each address in the first dataset, it can be assumed that there exists a match in the reference dataset. In the latter scenario, we have to provide for the case where there is no match for an address.

7.1   Linking with a Reference Dataset

This scenario usually occur during address cleansing. We deal with two address databases. The first database contains raw addresses, whereas the second database contains reference addresses. For each raw address, we search for its equivalent reference address, which provides a cleansed representation of the raw address.

In this experiment, we use two address databases:

- **AGA1** is a raw database collected by an Australian Government Agency. The database contains around 48 millions addresses most of which are Australian addresses. Addresses in this database are known to have significant data quality issues, with many incomplete and inaccurate addresses.
- **OpenAddress_Australia** contains more than 19 millions Australian addresses. All addresses are in standard form. This reference address database is open-source and can be downloaded from https://openaddresses.io. Almost all Australian addresses in AGA1 have a reference entry in OpenAddress_Australia.

We use the batch linkage algorithm to link addresses in AGA1 with addresses in OpenAddress_Australia. We extract order-preserving 2-word phrase tokens from the addresses and construct inverted indexes for both databases. We then compute character-based Jaccard index between each pair of shortlisted candidates. We accept a link if the Jaccard index exceeds a threshold τ.

Since we do not have a ground truth for the address cleansing result, we can not quantitatively assess the rate of false negatives (i.e. there exists a cleansed entry for a raw address but the algorithm cannot find it) in our linkage result. It is fair to say that essentially all data operations involving large databases have the same problem. It is therefore difficult to select the proper threshold value τ.

We propose the following mechanism for threshold selection. We implement address linkage with increasing thresholds, *e.g.*, $\{\tau_1 = 0.6, \tau_2 = 0.7, \tau_3 = 0.8\}$. We then use the result of the lowest threshold to benchmark that of higher thresholds for false negatives.

Figure 1 shows the percentage of true positives, false positives, and false negatives for the proposed method. These results are obtained by manually assessing 100 randomly sampled linked addresses. As we can see, when $\tau = 0.6$, which roughly requires a cleansed address to share 60% or more characters with the raw address, nearly 40% of raw addresses will find false cleansed forms. When τ increases to 0.7, the percentage of false positives drops to 12%. Conversely, 2% of raw addresses which used to find cleansed forms can no longer find them. This missing rate rises to 31% when τ increases to 0.8. Among the three values, $\tau = 0.7$ gives the best performance.

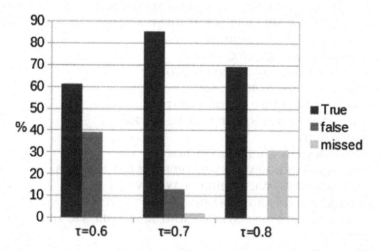

Fig. 1. The percentage of true positives (true), false positives (false), and false negatives (missed) of proposed addressing linkage algorithm with different thresholds. (Color figure online)

Table 1 lists 10 example links between AGA1 and OpenAddress_Australia found by our algorithm. Due to privacy concern, the addresses in these examples have been modified and does not reflect the original addresses. Besides the two linked addresses, we also provide the clean addresses found by Google Maps for AGA1 addresses. To protect anonymity we have encrypted the street names and suburb names. Interestingly, there are three addresses that Google Maps failed to process, yet were successfully linked by our algorithm. The fourth example in the table shows the limitation of using characters as tokens in the Jaccard index calculation. A simple tie-breaker postprocessing scheme using, for example edit distance, can be used to resolve such issues.

118   Y. Zhang et al.

Table 1. Address linkage between AGA1 and OpenAddress_Australia

	Address	Jaccard
AGA3	33 34-38 EHMNTV DIRTU NSW 5661	
Open	UNIT 33 34-38 EHMNTV STRET MOUNT DIRTU NSW 5661	1.66
Google	33/34-38 EHMNTV ST MOUNT DIRTU NSW 5661	
AGA3	53 741 ADGNR EFORST AKLR QLD 9368	
Open	UNIT 53 741 ADGNR AVENUE EFORST LAKE QLD 9168	1.60
Google	53/741 ADGNR AVE EFORST AEKL QLD 9168	
AGA3	972 4 CEOPRW LMOW NEW WALES 5133	
Open	UNIT 972 4 CEOPRW AFHRW ROADWAY LMOW NSW 5133	1.81
Google	NOT FOUND	
AGA3	713 311 AGHILNU AGHILNU ACT 5035	
Open	UNIT 731 311 AGHILNU PLACE AGHILNU ACT 5035	1.86
Open	UNIT 713 311 AGHILNU PLACE AGHILNU ACT 5035	1.86
Open	UNIT 317 311 AGHILNU PLACE AGHILNU ACT 5035	1.86
Google	713/311 AGHILNU PL AGHILNU ACT 5035	
AGA3	3 59 FGIS DEKNOR QLD QLD 9173	
Open	UNIT 3 59 FGIS STRET DEKNOR QLD 9173	1.81
Google	3/59 FGIS ST DEKNOR QLD 9173	
AGA3	9 NO 7 TO 2 CELMNT ADEGNO VIC 7362	
Open	UNIT 9 7-2 CELMNT STRET ADEGNO VIC 7362	1.82
Google	NOT FOUND	
AGA3	313 0 EGKNORW ABEHILTZ BAY 5133	
Open	UNIT 313 0 EGKNORW AVENUE ABEHILTZ BAY NSW 5133	1.83
Google	313/0 EGKNORW AVE ABEHILTZ BAY NSW 5133	
AGA3	MARGETIC 6 715 ABDFORST BELMNORU VIC 7123	
Open	FLAT 6 715 ABDFORST STRET HNORT BELMNORU VIC 7123	1.85
Google	NOT FOUND	
AGA3	43 3345 ACDHINSV AGRTV QLD 9355	
Open	UNIT 43 3345 ACDEHINSV ROAD MOUNT AGRTV EAST QLD 9355	1.64
Google	43/3345 ACDEHINSV RD MOUNT AGRTV EAST QLD 9355	
AGA3	78 03 ADELMNOR BELM VIC VIC 7133	
Open	UNIT 78 03 ADELMNOR STRET ACFORSTY VIC 7133	1.67
Google	78/03 ADELMNOR ST ACFORSTY VIC 7133	

7.2 Linking Two Arbitrary Datasets

This scenario occurs when people try to integrate two databases together. To test this scenario, we use two databases AGA1 and AGA2.

- **AGA2** contains around 18 millions addresses collected by a large Australian government department. Most addresses in AGA2 are Australian addresses. Addresses in this database may be incomplete and inaccurate. AGA1 and AGA2 are collected by different government agencies from different sources and for largely different original purposes.

We again use the batch linkage algorithm with 2-word phrase tokens for round 1 of Jaccard computations and character tokens for round 2. However, in this second address-linkage scenario, we can no longer use a simple threshold τ to reject false matches. This is because when linking with a reference dataset, if a street is included in the reference database, all individual addresses in this street are included. Therefore, if a raw address has a high score best match in the reference database, this best match is usually consistent with the raw address in every detail. However, in the scenario where we are linking two arbitrary databases, it is quite common for two databases to contain only two different addresses in the same street. These two addresses may have the highest matching score but still remain a false match. To complicate matters, a true match can also be a low score match due to data quality issues with both addresses.

One way to overcome this challenge is to require two matching addresses to have consistent numeric tokens. We say two sets of numeric tokens are consistent, if one set is a subset of the other.

We manually assess 100 randomly sampled AGA2 addresses. For each AGA2 address, we in order consider its top 3 matches in AGA1 database. If a match has consistent numeric tokens and is a true match, we label this AGA2 sample as true and no longer consider the remaining matches. If a match has consistent numeric tokens but is a false match, we label this AGA2 sample as false and no longer consider the remaining matches. If none of the top 3 matches has consistent numeric tokens with the query, this AGA2 sample is labelled as not found. Figure 2 shows the percentage of three labels in the 100 samples. It can be derived from Fig. 2 that, $59/(59 + 6) = 91\%$ of the samples are correctly linked.

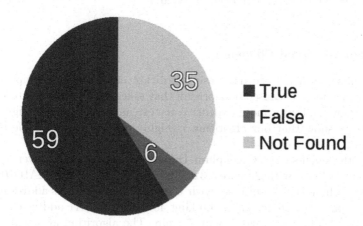

Fig. 2. Percentage of correctly linked (True), incorrectly linked (False), and not linked (Not Found) when joining AGA2 addresses to AGA1 addressees using proposed algorithm. (Color figure online)

Table 2 lists 10 example links between AGA1 and AGA2 found by our algorithm.

Table 2. Address linkage between AGA1 and AGA2

	Address	Jaccard
AGA1	4 EGNORU AEHLMT VIC 3095	
AGA2	4 EGNORU CRT AEHLMT NORTH VIC 3095	0.84
AGA1	528 LTUY HLU LTUY HLU QLD 4854	
AGA2	528 LTUY ADEHLSU RD LTUY QLD 4854	0.87
AGA1	45 EGHIMNS ADEGNO VIC 3175	
AGA2	RM 8 45 EGHIMNS ST ADEGNO VIC 3175	0.88
AGA1	6 EILS CEIMNRTY ABILRSUY DNOSW SA 5108	
AGA2	6 EILS CEIMNRTY RD ABILRSUY DNOSW SA 5108	0.96
AGA1	EL EILOSU 137 AILNT EFNRY EOV QUEN SLAND 4055	
AGA2	137 AILNT RD EFNRY EGORV QLD 4055	0.74
AGA1	80 ABEGLNRU DEHILS VI 3037	
AGA2	80 ABEGLNRU DR DEHILS VIC 3037	0.94
AGA1	141 ACEHLRS EHPRT 6005	
AGA2	141 ACEHLRS ST ESTW EHPRT WA 6005	0.80
AGA1	51 BENOR BELMNOT VIC 3216	
AGA2	2/51 BENOR DR BELMNOT VIC 3216	0.91
AGA1	97 ELOXY ABDPRUY WA 6025 TRA LIA	
AGA2	97 ELOXY AVE ABDPRUY WA 6025	0.87
AGA1	9 DLORS DENSY 2077	
AGA2	9 DLORS AVE AHIQSTU NSW 2077	0.68

7.3 Computational Efficiency

When dealing with a large database, algorithm efficiency is as important as algorithm accuracy, because an algorithm that takes days to finish is too expensive to deploy, and even more expensive to test under multiple configurations. Experiments show that our algorithm is highly efficient and scalable to large databases.

Using the open-source Greenplum Database running on 8 servers (1 master + 7 slaves), each with 20 cores, 320 GB, and 4.5 TB usable RAID10 space, linking 48 million AGA1 addresses with 13 million OpenAddress addresses using our algorithm takes about 5 min. Linking 48 million AGA1 addresses with 18 million AGA2 addresses takes about 7.5 min. The algorithm also scales essentially linearly in the number of servers in the Greenplum cluster dedicated to the task.

Note that the processing time of our algorithm depends more on the similarity between two databases than on the sizes of the two databases. The efficiency of the algorithm is due to the following factors:

1. The quantity of Jaccard index computation does not depend on the size of the database, but the number of addresses sharing common distinctive tokens.
2. Finding addresses sharing common distinctive tokens is done jointly for all addresses at the same time. This overhead does not depend on the number of addresses, but the number of distinctive tokens during the joining between two inverted indexes.

8   Parameters of the Algorithm

The use of Jaccard index to assess similarity between addresses in our algorithm is optional. Our implicit assumption is that there exists a function $d(x, y)$ which assesses the similarity between two addresses x and y. Blocking can reduce the number of evaluations of $d(x, y)$ without missing links, if $d(x, y) > \tau$ indicating x and y share a common token. In our two-round linkage, our implicit function is

$$d(x, y) = \begin{cases} J_{char}(x, y) & \text{if } J_{phrase}(x, y) > 0 \\ 0 & \text{otherwise} \end{cases} \tag{3}$$

One may design any other implicit function instead, replacing the Jaccard index with any other measurement.

The Jaccard index in round 2 of the comparison can also be replaced by almost any other similarity function, for example the Monge-Elkan function [13], which is suitable for addresses.

9   Conclusion

We have presented in this paper a novel address-linkage algorithm that:

1. links addresses as free text;
2. uses data-driven blocking keys;
3. extends the inverted index data structure to facilitate large-scale address linking;
4. is robust against data-quality issues; and
5. is practical and scalable.

The simplicity of the solution - a great virtue in large-scale industrial applications - may belie the slightly tortuous journey leading to its discovery; a journey laden with the corpses of a wide-range of seemingly good ideas like compressive sensing and other matrix factorisation and dimensionality-reduction techniques, nearest-neighbour algorithms like KD-trees, ElasticSearch with custom rescoring functions [8], rules-based expert systems, and implementation languages that range from low-level C, to R, Python, SQL and more. In retrospect, our algorithm can be interpreted as an application of a signature-based approach to efficiently compute set-similarity joins [1], where the abstract concept of sets is replaced with carefully considered set-representations of addresses, with a modern twist in its implementation on state-of-the-art parallel databases to lift the algorithm's scalability to potentially petabyte-sized datasets.

References

1. Arasu, A., Ganti, V., Kaushik, R.: Efficient exact set-similarity joins. In: Proceedings of the 32nd International Conference on Very Large Data Bases, pp. 918–929. VLDB Endowment (2006)
2. Cands, E.J., Romberg, J.K., Tao, T.: Stable signal recovery from incomplete and inaccurate measurements. Commun. Pure Appl. Math. **59**(8), 1207–1223 (2006)
3. Chang, L., Wang, Z., Ma, T., Jian, L., Ma, L., Goldshuv, A., Lonergan, L., Cohen, J., Welton, C., Sherry, G., Bhandarkar, M.: HAWQ: a massively parallel processing SQL engine in Hadoop. In: Proceedings of the 2014 ACM SIGMOD International Conference on Management of Data, pp. 1223–1234. ACM, New York (2014)
4. Christen, P.: Data Matching: Concepts and Techniques for Record Linkage, Entity Resolution, and Duplicate Detection. Springer, Heidelberg (2012). https://doi.org/10.1007/978-3-642-31164-2
5. Christen, P., Belacic, D.: Automated probabilistic address standardisation and verification. In: Australasian Data Mining Conference (AusDM05) (2005)
6. Christen, P., Churches, T., Hegland, M.: Febrl – a parallel open source data linkage system. In: Dai, H., Srikant, R., Zhang, C. (eds.) PAKDD 2004. LNCS (LNAI), vol. 3056, pp. 638–647. Springer, Heidelberg (2004). https://doi.org/10.1007/978-3-540-24775-3_75
7. Cohen, W.W., Ravikumar, P., Fienberg, S.E.: A comparison of string metrics for matching names and records. In: SIGKDD (2003)
8. Gormley, C., Tong, Z.: Elasticsearch: The Definitive Guide. O'Reilly Media, Sebastopol (2015)
9. Guo, H., Zhu, H., Guo, Z., Zhang, X., Su, Z.: Address standardization with latent semantic association. In: Proceedings of the 15th ACM SIGKDD International Conference on Knowledge Discovery and Data Mining, KDD 2009, pp. 1155–1164. ACM, New York (2009)
10. Halevy, A., Norvig, P., Pereira, F.: The unreasonable effectiveness of data. IEEE Intell. Syst. **24**, 8–12 (2009)
11. Hellerstein, J.M., Ré, C., Schoppmann, F., Wang, D.Z., Fratkin, E., Gorajek, A., Ng, K.S., Welton, C., Feng, X., Li, K., Kumar, A.: The MADlib analytics library or MAD skills, the SQL. PVLDB **5**(12), 1700–1711 (2012)
12. Kornacker, M., Behm, A., Bittorf, V., Bobrovytsky, T., Ching, C., Choi, A., Erickson, J., Grund, M., Hecht, D., Jacobs, M., Joshi, I., Kuff, L., Kumar, D., Leblang, A., Li, N., Pandis, I., Robinson, H., Rorke, D., Rus, S., Russell, J., Tsirogiannis, D., Wanderman-Milne, S., Yoder, M.: Impala: a modern, open-source SQL engine for Hadoop. In: CIDR (2015)
13. Monge, A., Elkan, C.: An efficient domain-independent algorithm for detecting approximately duplicate database records. In: DMKD 1997 (1997)
14. Zaharia, M., Chowdhury, M., Franklin, M.J., Shenker, S., Stoica, I.: Spark: cluster computing with working sets. In: Proceedings of the 2nd USENIX Conference on Hot Topics in Cloud Computing, HotCloud 2010, p. 10. USENIX Association, Berkeley (2010)

Time Series

SD-HOC: Seasonal Decomposition Algorithm for Mining Lagged Time Series

Irvan B. Arief-Ang$^{(\boxtimes)}$ (iD), Flora D. Salim, and Margaret Hamilton

Computer Science and Information Technology, School of Science,
Royal Melbourne Institute of Technology (RMIT), Melbourne, VIC 3000, Australia
{irvan.ariefang,flora.salim,margaret.hamilton}@rmit.edu.au

Abstract. Mining time series data is a difficult process due to the lag factor and different time of data arrival. In this paper, we present Seasonal Decomposition for Human Occupancy Counting (SD-HOC), a customised feature transformation decomposition, novel way to estimate the number of people within a closed space using only a single carbon dioxide sensor. SD-HOC integrates time lag and line of best fit model in the preprocessing algorithms. SD-HOC utilises seasonal-trend decomposition with moving average to transform the preprocessed data and for each trend, seasonal and irregular component, different regression algorithms are modelled to predict each respective human occupancy component value. Utilising M5 method linear regression for trend and irregular component and dynamic time warping for seasonal component, a set of the prediction value for each component was obtained. Zero pattern adjustment model is infused to increase the accuracy and finally, additive decomposition is used to reconstruct the prediction value. The accuracy results are compared with other data mining algorithms such as decision tree, multi-layer perceptron, Gaussian processes - radial basis function, support vector machine, random forest, naïve Bayes and support vector regression in two different locations that have different contexts.

Keywords: Ambient sensing · Building occupancy
Presence detection · Number estimation · Cross-space modeling
Contextual information · Human occupancy detection
Carbon dioxide · Machine learning

1 Introduction

Data mining technology is assimilated in human life and it helps solve many problems that could not be solved before. The problem we will consider in this paper is to do with building operational costs. From the U.S. Department of Energy, 35%–45% of the total operational costs within a building are spent on heating, ventilation, and air conditioning (HVAC) [1]. Due to this, substantial investment in the energy usage research area is needed to reduce HVAC costs in buildings based on their occupancy patterns. Reducing HVAC usage is

© Springer Nature Singapore Pte Ltd. 2018
Y. L. Boo et al. (Eds.): AusDM 2017, CCIS 845, pp. 125–143, 2018.
https://doi.org/10.1007/978-981-13-0292-3_8

equivalent to reducing the overall energy consumption. Furthermore, a Building Management System (BMS) can then intelligently adjust the HVAC based on the occupancy pattern so the comfort of the dwellers is not sacrificed.

Using sensor data to detect human's precence is the current trend in ambient sensing research area [2–6]. Yan higlighted the importance of occupant related research [7]. In [8], it was highlighted that carbon dioxide (CO_2) is the best ambient sensor predictor for detecting human presence. By using only CO_2, 91% accuracy was achieved for binary prediction, knowing the room is occupied or vacant [9] and have 15% accuracy for recognising the number of occupants. A hidden Markov model (HMM) was implemented for CO_2 dataset to predict human occupancy and 65%–80% range of accuracy was achieved for predicting up to 4 occupants [10].

In this paper, we propose a new algorithm for decomposing large datasets to extract the relevant features to be used for prediction and identification of seasonal trends. We can then apply the computations of these trends to various incomplete sets matching the time series to predict the relevant future features in the new dataset. We identify relevant seasonal trends in the data over time and apply these to the new dataset and use them to predict future trends in the data.

We have found this to be particularly useful for sensor data where we can extrapolate the CO_2 data to indoor human occupancy prediction with promising accuracy. We can match the sensor measurements for zero occupancy, at various times, possibly overnight and tune our predictions to optimise individual comfort and the overall carbon footprint of the building.

We apply our new feature transformation algorithm to the prediction of the number of people in a room at a particular time through the measurement of the carbon dioxide. Human occupancy prediction is an significant problem for the building industry because it enables the automation of heating, cooling and lighting systems. If it is known that certain rooms are empty or underutilised during certain times, operational costs and carbon footprint can be reduced with better planning and scheduling. When the rooms are not occupied, the building system can also adjust these facilities to keep the inhabitants comfortable. This framework is called seasonal decomposition for human occupancy counting (SD-HOC).

SD-HOC pre-processes the data and integrates various machine learning algorithms. The experiment is conducted on two different locations. There are two stages we have defined in our experiment. Firstly, SD-HOC result is compared with a variety of other data mining prediction algorithms such as decision tree, multi-layer perceptron, Gaussian processes - radial basis function, support vector machine and random forest. The second stage of our experiment compares SD-HOC with one of the best data mining prediction accuracy to predict the human occupancy number on different number of prediction days. There are three advantages of utilising SD-HOC:

1. It employs low equipment cost due to pre-installation;
2. SD-HOC ensures that users' privacy is protected;
3. It only uses CO_2 data, reducing the chance of errors caused by data integration.

The remainder of the paper is organised as follows. Section 2 presents the related work on current state-of-the-art indoor human occupancy methods. Section 3 covers the problem definition. Section 4 covers the features and Sect. 5 introduces SD-HOC framework. Section 6 describes experiments we conducted concerning set-up machine and multiple datasets. It also contains the results and comparisons with other data mining algorithms. Section 7 discusses the results and Sect. 8 concludes the paper with directions to the future work.

2   Background and Related Work

When using image processing techniques [11,12], the levels of accuracy for human occupancy detection can reach up to 80%. Unfortunately, these image processing methods raise privacy concerns. Research communities have been doing their best to propose various methods to detect human occupancy without using cameras or image processing.

We focus on utilising only CO_2 sensors alone and data to estimate the indoor human occupancy number. The main reason is because CO_2 sensors are already integrated with the BMS and ventilation infrastructure and are commonly installed in buildings.

Machine learning algorithms including a hidden markov model (HMM), neural networks (NN) and support vector machine latent (SVM latent) were implemented in [10] by using CO_2 data with the sensors deployed both inside and outside room. By feature engineering CO_2 data with first order and second order difference of CO_2, the accuracy achieved is between 65%–80%.

A mass balance approach was implemented to predict both human occupancy and occupant activity using CO_2 and door sensors in [13]. The authors mentioned that sources of error and uncertainty in this method are part of the limitation of this approach. CO_2 based occupancy detection in office and residential buildings was implemented in [14]. Binary occupancy accuracy prediction is 95.8% and the people counting accuracy is 80.6% for 2–3 person in each room.

PerCCS is a model with a non-negative matrix factorization method to count people [9] using only one predictor in CO_2. In predicting vacant occupancy, they achieved up to 91% but only 15% accuracy in predicting the number of occupants.

Overall, sensor-based detections have higher accuracy compared to radio-based detections. For example, Wi-Fi and RSSI signals achieved 63% accuracy for indoor detection with 9 occupants [15]. For occupancy counting, CO_2 sensors only have been experimented with the maximum of 42 occupants and accuracy limit of 15% [9]. A domain adaptation technique has been implemented for human occupancy counting with CO_2 and the prediction accuracy increases up to 12.29% compared to the baseline [16].

3   Problem Definition

Given significant motivations in our research, this paper presents the problem on how can we use data mining techniques and feature selection to predict the number of people by using a single CO_2 sensor. We would take the results to have similar accuracy to the state-of-the-art techniques in the occupancy detection field. In Fig. 1, the data shows there is a dependency between CO_2 and occupancy data.

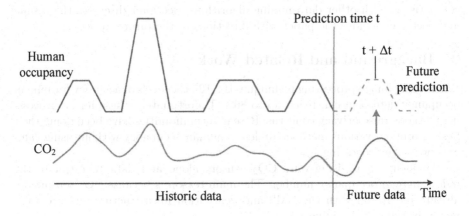

Fig. 1. Real-time prediction scenario for continuous t showing the amount of CO_2 fluctuations. The fundamental task is to predict the number of occupants at time $t + \Delta t$.

3.1 Scenario Assumption

Assume $|TS|$ represents the length of a time series, $TS = \{ts_1, ts_2, \ldots, ts_q\}$, where q means the number of sample points. In our time series datasets, we have two aspects:

– Carbon dioxide (CO_2) concentration C, defined as $C = \{C_1, C_2, \ldots, C_q\}$
– Indoor human occupancy O, defined as $O = \{O_1, O_2, \ldots, O_q\}$

Our framework only depends on CO_2 data set to calculate the prediction. This is where the challenge lies as the model needs to extract more features from a time series, which may seem simple, but it contains hidden trends. We introduce a term 'lagged time series' as a set of data in regression time series where each value relate to a situation in a surrounding context but does belong to different time frame.

3.2 Time Series

In time series prediction, analysing one-step-ahead prediction is different from analysing multi-step-ahead prediction. Predicting multi-step-ahead needs a more complex method due to the accumulation of errors and the number of uncertainties increasing with time. We focus on multi-step-ahead prediction with the support of one dependent variable to reduce uncertainties.

We have two different types of datasets: CO_2 concentration C and indoor human occupancy O. In order to explore the relationship between both factors above, we need to identify the relevant features by exploring the correlation between CO_2 concentration and indoor human occupancy and all of their decomposed components to find what correlations exist between each component.

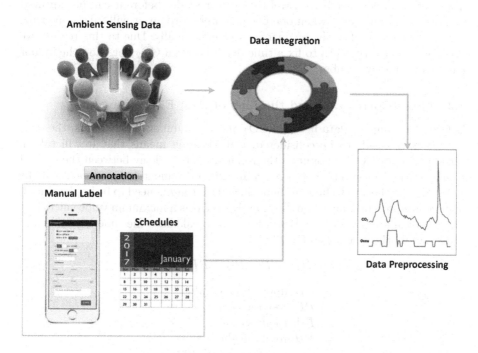

Fig. 2. Data collection and analysis framework.

4   The Features

This section explains our data pre-processing time series components, cross-correlation and line of best fit. Data pre-processing is crucial for our model as this step will further increase prediction analysis with various machine learning algorithms that we implemented in the experiment section. We collected data about both the CO_2 concentration from the sensor data and the number of humans in the room as shown in Fig. 2. Both data are pre-processed and integrated using our novel pre-processing method described below in the Subsect. 4.1. We transformed each data set using feature engineering into more features and applied our prediction model, SD-HOC, described in Sect. 5 to predict the indoor human occupancy.

4.1 Time Delay Components

Time delay issue is a problem because it takes time for the concentration of CO_2 to build up enough to measure a person. To model a real time delay we need the value of a time series regression function obtained after specific time lag. This issue normally happens in the majority of sensor data analyses as data obtained from sensor readers needs to travel to a sensor reader before it can be captured in storage. In our study, when one person enters a room, it will take some time before the CO_2 level in the air increases proportionally. Due to this reason, we must pre-process the data to fix a time delay between CO_2 data and the indoor human occupancy number.

4.2 Cross-Correlation and the Line of Best Fit

Before analysing the data between CO_2 and the number of occupants, the data lagging issue needs to be considered. Data lagging means that it will take a certain time for CO_2 to populate the room as there is delay between the time of people exiting (or entering) the room and the decrement (or increment) of the CO_2 value on the air. To find out how much data lagging need to be implemented, first we need to find upper bound value (UB). UB is a maximum value calculated based on the room volume. UB will be used to calculate the time lag value and is defined by the formula in Eq. 1.

$$UB = |(RL * RW * RH)/C| \qquad (1)$$

UB upper bound value
RL room length
RW room width
RH room height
C constant value (100)

For each dataset from 0 min time lag to UB minutes time lag, the correlation of CO_2 data with the number of occupancies is measured. If the room size is small, the UB value will be 1. The larger is room is, the bigger the UB value is. In our case study, for the small room A, the UB value will be 1 and for big room B, the upper bound of N is 60. This value is aligned with the explanation above due to the large size of the big room B.

To calculate a line of best fit, we need to calculate the slope value between CO_2 and occupancy data, defined by Eq. 2.

$$SL = \frac{\sum(O_t - \bar{O}_t)(C_t - \bar{C}_t)}{\sum(O_t - \bar{O}_t)^2} \qquad (2)$$

SL slope of the linear regression line
O_t occupancy value
$\bar{O}_t$ sample means of the known occupancy value
C_t CO_2 value
$\bar{C}_t$ sample means of the known CO_2 value

Next, the intercept value between both data sets needs to be calculated using the formula in Eq. 3.

$$IC = \bar{C}_t - SL * \bar{O}_t \tag{3}$$

IC intercept of the linear regression line
$\bar{C}_t$ sample means of the known CO_2 value
$\bar{O}_t$ sample means of the known occupancy value

The main formula for the line of best fit (LBF) is shown in Eq. 4.

$$LBF = (O_t - (SL * C_t + IC))^2 \tag{4}$$

O_t occupancy value
SL slope of the linear regression line
C_t CO_2 value
IC intercept of the linear regression line

4.3   Time Lag

For each line of best fit from Subsect. 4.2, we calculate the mean squared error (MSE), root-mean-square deviation (RMSD) and the normalised root mean squared error (NRMSE). The formula for calculating NRMSE is shown in Eq. 5.

$$NRMSE = \frac{\sqrt{\frac{1}{n} \sum_{t=1}^{n} (C_t - \bar{C}_i)^2}}{O_{max} - O_{min}} \tag{5}$$

$NRMSE$ normalized root mean square error
t total number of data set
C_t CO_2 value
$\bar{C}_t$ sample means of the known CO_2 value
O_{max} maximum occupancy value
O_{min} minimum occupancy value

This step is repeated UB times for each time lag. For time lag analysis, we use least square regression to compare each NRMSE from time lag 0 until time lag UB. We pick the lowest number of NRMSE value as our time lag value (TL). The TL value formula is shown in Eq. 6 and it performs as our baseline time lag for the data analysis.

$$TL = min(NRMSE) \tag{6}$$

For the academic staff room, the TL value is 0. This value represents no time lag is needed for this analysis. For the cinema theatre, the lowest number of error value happens at time lag TL = 32 as shown in Fig. 3. This TL value is our base for the cinema theatre data analysis. So for the entire cinema data analysis process, we use time lag 32.

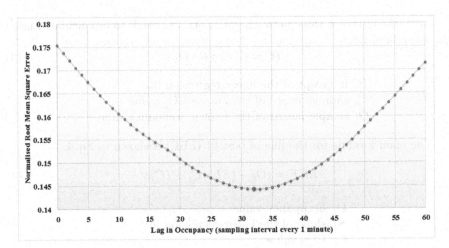

Fig. 3. Ordinary Least Square Regression Normalised Root Mean Squared Error (NRMSE) between CO_2 data and actual occupancy for 60 min time lag.

5   The Framework

There is no linear relationship between CO_2 and indoor human occupancy. For this reason, we introduce a new SD-HOC analysis framework in Fig. 2 to address this non-linear correlation issue by decomposing both CO_2 and occupancy data shown in Fig. 4. In this paper, for the main decomposition method we are using is known as seasonal trend decomposition (STD).

The core feature transformation prediction model will be explained in the next following subsections. The first subsection discusses STD in detail. The next subsection explains the correlation model for trend, seasonal and irregular features. The last subsection presents zero pattern adjustment (ZPA), a new method for analysing conditions when the room is vacant. ZPA method can increase the overall accuracy. This model needs to be re-trained for different locations to obtain the most optimal accuracy results.

5.1   Seasonal-Trend Decomposition

STD is a decomposition technique in time series analysis. X-11 method with moving average is one of the most famous variants [17] and X12-ARIMA is the most recent variant [18]. STD is an integral part of our framework.

To understand each time series data, we utilise STD to decompose the data into four main features: trend, cyclical, seasonal and irregular. The trend feature (T_t) represents the long-term progression of the time series during its secular variation. The cyclical feature (C_t) reflects a repeated but non-periodic fluctuation during a long period of time. The seasonal feature (S_t) is a systematic and regularly repeated event during short period of time. And the irregular feature (e_t also known as error or residual) is a short term fluctuation from the time series and is the reminder after the trend, cyclical and season features have been removed.

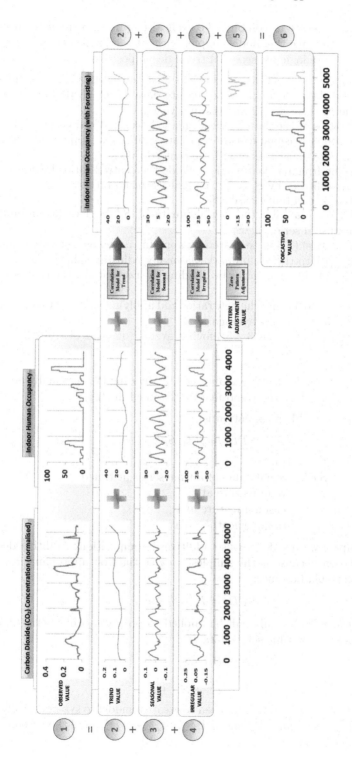

Fig. 4. Seasonal decomposition for human occupancy counting (SD-HOC) analysis framework.

In this paper, we decide to combine the cyclical feature into trend feature due to its similarity to make the model simpler without sacrificing the accuracy.

Below is the core logic for seasonal trend decomposition:

1. Calculate 2×12 moving average in the raw data (both CO_2 and occupancy datasets) to obtain a rough trend feature data T_t for all period (12 is the default due to there are 12 months in a year).
2. Calculate ratios of the data to trend, named "centred ratios" (y_t/T_t).
3. To form a rough seasonal feature (S_t) data estimation, apply separate 2×2 moving average to each month of the centred ratios.
4. To obtain the irregular feature (e_t), divide the centred ratios by S_t.
5. Multiply modified e_t by S_t to get modified centred ratios.
6. Repeat step 3 to obtain revised S_t.
7. Divide the raw data by the new estimate of S_t to give the preliminary seasonal adjusted series, y_t/S_t.
8. The trend feature (T_t) is estimated by applying a weighted Henderson moving average [19] to the preliminary seasonally adjusted values.
9. Repeat step 2 to get new ratios by dividing the raw data by the new estimate of T_t.
10. Repeat Steps 3 to 5 using the new ratios and applying a 3×5 moving average instead of a 3×3 moving average.
11. Repeat step 6 but using 3×5 moving average instead of a 3×3 moving average.
12. Repeat step 7.
13. Finally the reminder feature is obtained by dividing the seasonally adjusted data from step 12 by the trend feature obtained in step 8.

Our customised STD formulation is:

$$STD_t = f(T_t, S_t, e_t) \tag{7}$$

t time
STD_t actual value of a time series at time t
T_t trend feature at t
S_t seasonal feature at t
e_t irregular feature at t

In this paper, we decided to use additive decomposition. Additive decomposition is chosen because is the simplest to give the first approximation. Our overall STD formula becomes:

$$STD_t = T_t + S_t + e_t \tag{8}$$

This general STD formula will be applied to both CO_2 time series dataset and human occupancy time series datasets:

$$C_t = T_t^C + S_t^C + e_t^C \tag{9}$$

$$O_t = T_t^O + S_t^O + e_t^O \tag{10}$$

To predict O_{t+1} up to O_{t+n}, we need to create a model to systematically predict each of T_{t+1}^O, S_{t+1}^O and e_{t+1}^O up to T_{t+n}^O, S_{t+n}^O and e_{t+n}^O and then reconstruct the new prediction dataset using additive method.

5.2   Correlation Models

There are three correlation models for features of trend, seasonal and irregular in the following subsections.

Correlation Model for Trend Feature (T_t). The definition for the trend feature (T_t) is the long-term non-periodic progression of the time series during its secular variation. Due to this, we assume that the trend feature for the CO_2 dataset (T_t^C) will be similar to the trend feature for indoor human occupancy (T_t^O) because there is dependency between both dataset.

Correlation model for trend feature start with checking the similarity between both trend features. We use Pearson product-moment Correlation Coefficient (PCC) to validate it as shown below:

$$r = \frac{n(\sum xy) - (\sum x)(\sum y)}{\sqrt{[n\sum x^2 - (\sum x)^2][n\sum y^2 - (\sum y)^2]}} \tag{11}$$

$$
\begin{aligned}
&r \text{ correlation coefficient} \\
&x \text{ dataset x} \\
&y \text{ dataset y} \\
&n \text{ number of sample points}
\end{aligned}
\tag{12}
$$

Correlation coefficient Pearson's r value ranges from -1 to $+1$. If the value is >0.7, the correlation between both datasets is strongly positive. If the correlation is less than 0.7, data pre-processing needs to be redone to find the new TL value (Eq. 6).

Once it passes the validation step, polynomial M5 linear regression is implemented. We chose the M5 method because it will build trees whose leaves are associated with multivariate linear models and the nodes of the tree are chosen over attributes that maximise the expected error reduction, given by the Akaike Information Criterion (AIC). AIC is a measure to check the relative goodness of fit of a statistical model [20]. The purpose of using AIC is to evaluate the model. The value for each of trend feature needs to be a positive value so we put the absolute value on both the CO_2 ($|T_t^C|$) and human occupancy trend features ($|T_t^O|$). The main formula for trend feature correlation is shown below:

$$|T_t^O| = |\alpha_0 + \alpha_1(T_t^C) + \alpha_2(T_t^C)^2 + \ldots + \alpha_n(T_t^C)^n + \epsilon| \tag{13}$$

Linear regression with M5 will output each α_n and ϵ value. With these parameters, the future trend for T_{t+n}^O can be obtained.

Correlation Model for Seasonal Feature (S_t). The seasonal feature (S_t) is a systematic and regularly repeated event during short period of time. Due to this characteristic, every seasonal feature can be fitted by a finite Fourier series. To correlate S_t^C and S_t^O, we use Dynamic Time Warping (DTW), a pattern matching technique to score the similarity between the shape of specific signal within certain duration [21]. The full correlation algorithm is implemented in Algorithm 1 to find regularly repeated events within each S_t.

Algorithm 1. Finding a repeated event inside seasonal feature

---

1: **procedure** REPEATED_EVENT(S_t)
2:    $s_t^{temp}, s_t^{fin} \subset S_t$
3:    $len \leftarrow 0$ ▷ len: Length for s_t^{temp}
4:    $a \leftarrow S_t[len]$ ▷ a: Start Point
5:    **for** each node $i \in S_t$ **do**
6:        $len++$
7:        $s_t^{temp} \leftarrow s_t^{temp} + S_t[i]$
8:        **if** $a = S_t[i]$ **then**
9:            **if** DTW($s_t^{temp}, S_t[i+1..i+len]$) > 95 **then**
10:                $s_t^{fin} \leftarrow s_t^{temp}$
11:                **break**
12:            **end if**
13:        **end if**
14:    **end for**
15:    **return** s_t^{fin}
16: **end procedure**

Once we find repeated event in s_t^{fin} for both the CO_2 and occupancy seasonal features, we compare the length of $s_t^{fin(O)}$ and $s_t^{fin(C)}$. If the length of $s_t^{fin(O)} < s_t^{fin(C)}$, we apply an interpolation method inside $s_t^{fin(O)}$ so both have the same length. If the length of $s_t^{fin(O)} > s_t^{fin(C)}$, we apply data reduction method so finally both have the same length. The final regression equation for seasonal feature correlation is shown below:

Correlation Model for Irregular Feature (e_t). Due to similar characteristics between trend and irregular features, we apply the same correlation method from the trend feature:

$$|e_t^O| = |\beta_0 + \beta_1(e_t^C) + \beta_2(e_t^C)^2 + \ldots + \beta_n(e_t^C)^n + \gamma| \tag{14}$$

The only difference from the trend feature is that we do not need to validate it using PCC as the shape of the irregular feature will depend more on its trend and seasonal features.

5.3   Zero Pattern Adjustment

In human occupancy prediction research, inferring knowledge when a room is vacant is paramount. By minimising false positives, the accuracy prediction can be improved. The Zero pattern adjustment (ZPA) method learns the behaviour from previous historical data and makes some smart adjustments for a vacant room when the normal algorithm returns incorrect prediction. The ZPA technique overlays all previous dataset and puts them on a single 24-h x-axis chart to determine the earliest start and end points when the room is vacant each day during the night to dawn period. We symbolise ZPA as zpa_t^O.

For our main occupancy model, we integrate each feature to get the occupancy prediction value.

6   Experiments and Results

In this section, our model is assessed for two different locations with distinct contexts to ensure the model's adaptability to various conditions. The first location is a small room A, belonging to one staff member at RMIT University, Australia. This room is chosen for human occupancy prediction since a controlled experiment can be conducted for an extended period of data collection.

The second dataset was collected inside a cinema theatre in Mainz, Germany [22]. Cinema theatre is chosen as another setting due its nature of having fluctuating numbers of people throughout the day. The numbers of people in the audiences can reach hundreds and can decrease to zero within a few hours. We will address this room as big room B.

6.1   Experiment Setting

Small Room A. We use a commercial off-the-shelf Netatmo urban weather station(Range: 0–5000 ppm, accuracy: ± 50 ppm) to read and collect ambient CO_2 data. The experiment took place between May and June 2015. The dataset is uploaded to a cloud service for integration purposes. We selected two weeks data from the whole dataset and used them in the further analysis. The room size is 3×4 m.

Big Room B. The cinema dataset were collected between December 2013 and January 2014 [22]. The dataset was collected using mass spectrometry machinery installed on the air ventilation system. The air flows from the screening room via the ventilation system to the mass spectrometer for data analysis.

Experiment Tool. We utilised WEKA, MATLAB and R to help us perform this experiment. WEKA is used for polynomial linear regression with M5 method for both correlation models for trend and irregular features (Subsect. 5.2). We also used WEKA for majority of data mining algorithms such as multi-layer perceptron, Gaussian processes (with kernel RBF), support vector machine, random forest, naïve Bayes, decision tree (with random tree) and decision tree (with M5P). MATLAB code is run for the baseline method, SVR and its prediction result. We used R to integrate all the data, including decomposition of STD and the majority of data pre-processing.

6.2   Experiment Parameters

SD-HOC model predicts each future value for the whole period of time based on specific time window. To understand this model better and how well it performs compared with the baseline, we define x, accuracy error tolerance parameter. Zero units error tolerance means only the exact number recognised is considered as true positive. For example with ten units error tolerance, if the real indoor human occupancy is 150 people, the prediction shows as low as 140 or as high as

138   I. B. Arief-Ang et al.

160 is considered correct as it is within ±10 units error tolerance. The parameter x value will be different based on the size of the room.

Each machine learning algorithms data has been preprocessed using the same method as in Sect. 4 to ensure that the comparison is fair.

Experiment for Small Room a Dataset. For an academic staff room dataset, we used 5-min time window. Total data that we gathered from this room are 4,019 data spread in 14 days. Due to the small room size, we decided not to use time lag for data analysis as there is a negligible period between exhaling process and sensor reading. For this room, we have seven pairs of the training-test dataset. It starts with seven days of training dataset and seven days of test dataset. It ends with 13 days of training dataset to predict one-day test dataset.

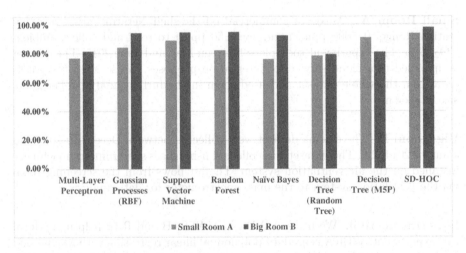

Fig. 5. Accuracy result of various machine learning algorithms.

Experiment for Big Room B Dataset. For cinema theatre dataset, we use 3-min time window for data analysis. Data that we gathered from this cinema theatre consisting of 68,640 instances spread over 23 days. The cinema theatre capacity is up to 300 people and for this experiment, we run the line of best fit for time lag 0 to time lag 60. The lowest NRMSE is at time lag 32 and we use time lag 32 as time lag baseline. This time lag is appropriate as bigger room need larger time lag for the model to have a better accuracy. For this room, we decided to use December 2013 data for training and January 2014 data for testing. Then we replicated it in the similar method by giving one day from testing dataset to training dataset and ran the model again. This method is repeated until the test dataset only consisted of one day of data.

6.3 Experiment Results with Other Data Mining Algorithms

From Table 1 and Fig. 5, we run each data mining algorithms and compare the result with our novel model, SD-HOC. SD-HOC have the highest accuracy prediction with 93.71% accuracy for staff room and 97.73% for cinema theatre.

Table 1. Accuracy result of various machine learning algorithms.

Machine learning	Small Room A	Big Room B
Multi-layer perceptron	76.69%	81.39%
Gaussian processes (RBF)	84.09%	94.09%
Support vector machine	88.86%	94.55%
Random forest	82.16%	94.75%
Naïve Bayes	75.89%	92.40%
Decision tree (Random tree)	78.23%	79.26%
Decision tree (M5P)	90.87%	80.68%
SD-HOC	**93.71%**	**97.73%**

6.4 Experiment Result with SVR on Different Number of Prediction Days

Support vector regression (SVR) have the highest prediction accuracy compared to the other data mining algorithms. Due to this reason, we run the experiment with the different training and testing to compare SD-HOC and the state-of-the-art machine learning algorithm baseline, SVR.

Evaluation and Baseline. To evaluate the result, we divide the data into 2 equal parts. The first part is the training dataset and the second one is the test dataset. To be able to understand how well the model fits for a longer duration, we repeat the division of training and test dataset by adding one day data from the test dataset to the training dataset. This replication is repeated again until the test dataset has only one single day and the rest belong to training dataset. This incremental days of training and reduction in testing evaluation method ensure the robustness of model.

Experiment Result for Small Room A Dataset. From Fig. 6, SD-HOC performed better than the baseline on average by 4.33%. As the last two days are Saturday and Sunday, both SD-HOC and the baseline models correctly predict zero occupancy for each day. In fact, they are vacant for the whole day.

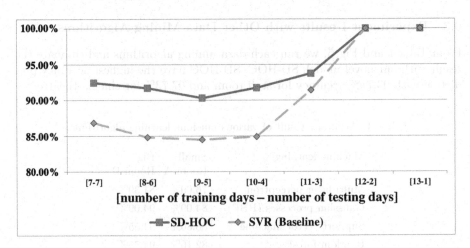

Fig. 6. Small room A dataset - The comparison for indoor human occupancy.

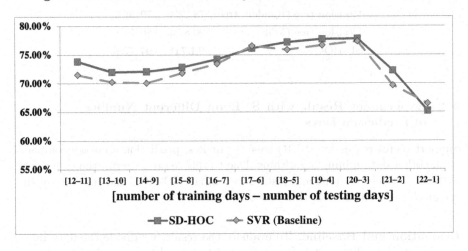

Fig. 7. Big room B dataset - The comparison for indoor human occupancy.

Experiment Result for Big Room B Dataset. For the cinema dataset, the comparison accuracy result is shown in Fig. 7. SD-HOC method performed better than the baseline method and on average SD-HOC method has 8.5% higher accuracy in predicting indoor human occupancy. The highest prediction accuracy was found when we used 22 days data for training to predict the number of human occupants the next day.

The results from Fig. 7 show that SD-HOC method is more accurate in predicting indoor human occupancy. This result is encouraging. Furthermore, we can observe that the accuracy for less number of days prediction is higher than for more days prediction, which is aligned with the results from academic staff room experiment.

7   Discussion

Our experiment shows that our new framework has the highest accuracy than most of data mining algorithms for both small and large rooms as shown in Fig. 5. Furthermore, this SD-HOC model is robust enough to handle different scales of data, proven by performing evaluation of our proposed model in two environments with different contexts such as room size and maximum number of occupants.

Compared with the baseline, SVR, our framework shows a better prediction accuracy over differing numbers of days for both the training and testing periods. This result demonstrates that seasonal decomposition can be utilised for predicting indoor human occupancy. The SD-HOC model can be used in many applications and is not limited to human occupancy prediction as it is based on seasonal decomposition methods.

SD-HOC performs well in comparison to other machine learning algorithms due to the feature transformation step, where, for each transformed feature, a set of relevant algorithms is run. For the small room A, SVM and decision tree method are the next best in prediction accuracy after SD-HOC. It is due to the fact that the number of people in this room fluctuated less from hour to an hour. There was also a stable CO_2 concentration for an extended period.

For the big room B, SVR and random forest are the next best in prediction accuracy after SD-HOC. Random forest behaves well when irrelevant features are present or these features have skewed distributions. The number of people in the big room could fluctuate from zero, or a vacant room to hundreds of people within 10–15 min. The SVM technique enables accurate discrete categorical labels to be predicted. This is why SVR is chosen as a baseline in Sect. 6.4.

8   Conclusion, Limitation and Future Work

Data mining algorithm roles in human life are becoming more important and its technology can be assimilated in human daily life. SD-HOC utilises several data mining algorithms and contributes to building and room occupancy counting. By understanding and knowing the numbers of people within a building, the heating, cooling, lighting control, building energy consumption, emergency evacuation, security monitoring and room utilisation can be made more efficient.

Although research in the human occupancy area has been studied with various methods including the use of ambient sensors, occupancy models that have been studied in previous work require the use of many sensors. In this experiment, we use a single sensor that is commonly available in the BMS to reduce the cost and complexity as more sensors can mean less reliability.

There are many possibilities that can be explored by using this technique. SD-HOC can be used for any time series dataset to predict another time series dataset as long as there is some dependency between those two data sets. SD-HOC is more than a simple correlation model and can solve many problems that a simple correlation model will not be able to solve.

CO_2 that is generated by human beings is affected by levels of physical activity. These different levels of activity such as walking, standing, or sitting could produce distinct CO_2 concentrations for the same individual. Also, the CO_2 rate in nature fluctuates around the day, reaching a higher value during the noon and dipping to a lower value at midnight. These CO_2 related facts could be integrated into the future works.

As our research was focussed on two locations and datasets, we plan to extend this research to other places that have different environmental dynamics and characteristics. For future work, other decomposition models and real-time online learning can be pursued to enhance the performance. Furthermore, from our research, indoor human occupancy could be related to certain events like public holidays and hence this feature could be included in future studies.

Acknowledgements. The authors would like to thank Joerg Wicker from University of Mainz for providing the cinema dataset used in this paper. This research is supported by the Australian Government Research Training Program Scholarship and two RMIT and Siemens Sustainable Urban Precinct Project (SUPP) grants: "iCo2mmunity: Personal and Community Monitoring for University-wide Engagement towards Greener, Healthier, and more Productive Living" and "The Greener Office and Classroom".

References

1. U.S. Department of Energy (DOE): Building Energy Databook. Technical report (2010)
2. Candanedo, L.M., Feldheim, V.: Accurate occupancy detection of an office room from light, temperature, humidity and CO_2 measurements using statistical learning models. Energy Build. **112**, 28–39 (2016)
3. Ekwevugbe, T., Brown, N., Pakka, V.: Realt-time building occupancy sensing for supporting demand driven hvac operations. Energy Systems Laboratory (2013)
4. Hailemariam, E., Goldstein, R., Attar, R., Khan, A.: Real-time occupancy detection using decision trees with multiple sensor types. In: Proceedings of the 2011 Symposium on Simulation for Architecture and Urban Design, pp. 141–148. Society for Computer Simulation International (2011)
5. Khan, A., Nicholson, J., Mellor, S., Jackson, D., Ladha, K., Ladha, C., Hand, J., Clarke, J., Olivier, P., Plötz, T.: Occupancy monitoring using environmental & context sensors and a hierarchical analysis framework. In: BuildSys@ SenSys, pp. 90–99 (2014)
6. Leephakpreeda, T.: Adaptive occupancy-based lighting control via grey prediction. Build. Environ. **40**(7), 881–886 (2005)
7. Yan, D., OBrien, W., Hong, T., Feng, X., Gunay, H.B., Tahmasebi, F., Mahdavi, A.: Occupant behavior modeling for building performance simulation: current state and future challenges. Energy Buildings **107**, 264–278 (2015)
8. Ang, I.B.A., Salim, F.D., Hamilton, M.: Human occupancy recognition with multivariate ambient sensors. In: 2016 IEEE International Conference on Pervasive Computing and Communication Workshops (PerCom Workshops), pp. 1–6. IEEE (2016)
9. Basu, C., Koehler, C., Das, K., Dey, A.K.: PerCCS: person-count from carbon dioxide using sparse non-negative matrix factorization. In: Proceedings of the 2015 ACM International Joint Conference on Pervasive and Ubiquitous Computing, pp. 987–998. ACM (2015)

10. Lam, K.P., Höynck, M., Dong, B., Andrews, B., Chiou, Y.S., Zhang, R., Benitez, D., Choi, J., et al.: Occupancy detection through an extensive environmental sensor network in an open-plan office building. IBPSA Building Simul. **145**, 1452–1459 (2009)
11. Erickson, V.L., Lin, Y., Kamthe, A., Brahme, R., Surana, A., Cerpa, A.E., Sohn, M.D., Narayanan, S.: Energy efficient building environment control strategies using real-time occupancy measurements. In: Proceedings of the First ACM Workshop on Embedded Sensing Systems for Energy-Efficiency in Buildings, pp. 19–24. ACM (2009)
12. Lee, H., Wu, C., Aghajan, H.: Vision-based user-centric light control for smart environments. Pervasive Mob. Comput. **7**(2), 223–240 (2011)
13. Dedesko, S., Stephens, B., Gilbert, J.A., Siegel, J.A.: Methods to assess human occupancy and occupant activity in hospital patient rooms. Build. Environ. **90**, 136–145 (2015)
14. Cali, D., Matthes, P., Huchtemann, K., Streblow, R., Müller, D.: CO_2 based occupancy detection algorithm: experimental analysis and validation for office and residential buildings. Build. Environ. **86**, 39–49 (2015)
15. Depatla, S., Muralidharan, A., Mostofi, Y.: Occupancy estimation using only WIFI power measurements. IEEE J. Sel. Areas Commun. **33**(7), 1381–1393 (2015)
16. Arief-Ang, I.B., Salim, F.D., Hamilton, M.: DA-HOC: semi-supervised domain adaptation for room occupancy prediction using CO_2 sensor data. In: Proceedings of the 4th ACM International Conference on Systems for Energy-Efficient Built Environments (BuildSys 2017), pp. 1–10. ACM (2017)
17. Shiskin, J., Young, A.H., Musgrave, J.C.: The X-11 variant of the census method II seasonal adjustment program. Number 15. US Department of Commerce, Bureau of the Census (1965)
18. Findley, D.F., Monsell, B.C., Bell, W.R., Otto, M.C., Chen, B.C.: New capabilities and methods of the X-12-ARIMA seasonal-adjustment program. J. Bus. Econ. Stat. **16**(2), 127–152 (1998)
19. Hyndman, R.J.: Moving Averages, pp. 866–869. Springer, Heidelberg (2011)
20. Akaike, H.: A new look at the statistical model identification. IEEE Trans. Autom. Control **19**(6), 716–723 (1974)
21. Petitjean, F., Ketterlin, A., Gançarski, P.: A global averaging method for dynamic time warping, with applications to clustering. Pattern Recogn. **44**(3), 678–693 (2011)
22. Wicker, J., Krauter, N., Derstorff, B., Stönner, C., Bourtsoukidis, E., Klüpfel, T., Williams, J., Kramer, S.: Cinema data mining: the smell of fear. In: Proceedings of the 21th ACM SIGKDD International Conference on Knowledge Discovery and Data Mining, pp. 1295–1304. ACM (2015)

An Incremental Anytime Algorithm for Mining T-Patterns from Event Streams

Keith Johnson$^{(\boxtimes)}$ (iD) and Wei Liu

School of Computer Science and Software Engineering,
The University of Western Australia, Crawley, Australia
keiflager@gmail.com, wei.liu@uwa.edu.au

Abstract. Temporal patterns that capture frequent time differences occurring between items in a sequence are gaining increasing attention as a growing research area. Time-interval sequential patterns (also known as T-Patterns) not only capture the order of symbols but also the time delay between symbols, where the time delay is specified as a time-interval between a pair of symbols. Such patterns have been shown to be present in many different types of data (e.g. spike data, smart home activity, DNA sequences, human and animal behaviour analysis and the like) which cannot be captured by other pattern types. Recently, several mining algorithms have been proposed to mine such patterns from either transaction databases or static sequences of time-stamped events. However, they are not capable of online mining from streams of time-stamped events (i.e. event streams). An increasingly common form of data, event streams bring more challenges as they are often unsegmented and with unobtainable total size. In this paper, we propose a mining algorithm that discovers time-interval patterns online, from event streams and demonstrate its capability on a benchmark synthetic dataset.

1 Introduction

A massive amount of data is being produced in continual streams, in real time. Event streams—streams that contain time-stamped, instantaneous events are of particular importance and exist in many situations such as contact sensors in a smart home, social media activities, web traffic logs, weather sensors, neuron spike recordings and the like. Mining temporal patterns from such event streams are of immense value to applications such as activity recognition, human and animal behaviour analysis, weather pattern detection and understanding of neuron spike patterns. Traditional data mining algorithms ignore the time difference between events, instead they focus only on the sequential ordering of the events, notably in Frequent Sequential Pattern (FSP) mining (Agrawal and Srikant 1995). Recent work by Chen et al. (2003) and Hu et al. (2009) extend FSP algorithms to mine time-interval sequential patterns which not only capture the order of symbols but the time differences between them. These approaches mine time-interval patterns from databases of time-stamped sequence transactions or static sequences of time-stamped events.

© Springer Nature Singapore Pte Ltd. 2018
Y. L. Boo et al. (Eds.): AusDM 2017, CCIS 845, pp. 144–157, 2018.
https://doi.org/10.1007/978-981-13-0292-3_9

In human and animal behaviour analysis a type of time-interval pattern, known as T-Patterns, introduced by Magnusson (2000), capture time-interval patterns in a hierarchical tree structure, shown in Fig. 1, and are built in a bottom-up hierarchical fashion. One advantage is that as new patterns are discovered (e.g. AB in Fig. 1), their occurrences are represented using new symbols. These new symbols are populated back into the dataset such that they are included as input for subsequent mining. In this way more complex T-Pattern structures are built iteratively by combining simpler ones. Another advantage is that simpler substructures can be shared across multiple T-Patterns.

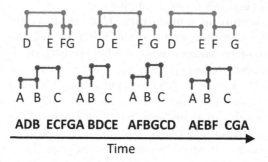

Fig. 1. An example stream of events and the occurrences of two T-Patterns in the input. The black letters represent the event stream input. The first hierarchical T-Pattern (blue, 4 occurrences) represents sequence ABC with time-interval constraints between AB and BC. The second T-Pattern (green, 3 occurrences) represents sequence DEFG with 3 time-interval constraints DE, FG and DF. (Color figure online)

In this paper we adapt Magnusson's definition (Magnusson 2000) to discover T-Patterns, *online*, from *event streams*. We propose an anytime algorithm for mining T-Patterns that clusters samples taken from a (limited) history of events to generate candidate time-intervals. Candidates are then evaluated with a heuristic score. This mining occurs iteratively, producing just one T-Pattern per instance of the algorithm—an anytime algorithm (run at a user specific frequency) that performs repeated incomplete searches rather than one complete search over a whole dataset. The overall goal of our approach differs slightly from the existing approaches, due to the different nature of the data (i.e. unsegmented and streaming) and the envisioned types of applications. The motivating application of our approach is to learn and continually update a model (in this case, a set of discovered T-Patterns), online, to support an online decision making process e.g. an agent embedded in a mobile robot, responsive smart home system or network traffic anomaly detection. Thus the primary goals of the proposed mining algorithm are to run as an anytime algorithm (to be suitable for online learning and with adjustable parameters to control computation time) and to mine fewer but high quality patterns (for model efficiency). Quality is hard to define without knowing the application, however we employ heuristics for this purpose.

The paper is structured as follows. Section 2 reviews the relevant literature on time-interval patterns and mining techniques. We describe our proposed mining algorithm in Sect. 3. Section 4 explains the experimental setup and provide results for a benchmark synthetic dataset. The paper concludes in Sect. 5 with an outlook to future work.

2   Literature Review

Here we review methods from two different camps; the data mining community and the human/animal behaviour community.

From the **data mining community**, Chen et al. (2003) introduce time-interval sequential patterns, that extend FSPs by defining a time-interval between every pair of successive symbols within the pattern. They define a time-interval sequential pattern as $(x_1, i_1, x_2, i_2, \ldots, x_n)$ where x_j is an occurrence of a symbol $s \in S$, the set of all symbols, and i_j is a time-interval between the occurrence of x_{j-1} and x_j, which is from the set of all possible time-intervals $ti \in TI$. They propose two algorithms, based on the conventional Apriori algorithm and the PrefixSpan algorithm, that mine frequent patterns of this type from a database of time-stamped sequence transactions. The mining algorithms aim to mine the maximum number of time-interval sequences that meet a minimum support threshold. Discovering appropriate time-intervals is one of the difficult challenges of mining this pattern type. In the algorithm by Chen et al. (2003), the mined time-intervals are constrained to a set of 3, 5 or 7 possible time-intervals that are determined at the beginning of the algorithm, based on the data. The proposed methods are evaluated on a synthetic dataset of transaction sequences adapted from a benchmark FSP dataset (Agrawal and Srikant 1995).

In time-interval sequential patterns, the time-intervals capture information about the timing between successive pairs within a pattern but not between more distant pairs. Hu et al. (2009) address this with multi-time-interval sequential patterns that extend (single) time-interval sequential patterns by defining time-intervals between all pairs of symbols in the pattern. This pattern is defined as $(x_1, I_1, x_2, I_2, \ldots, x_n)$ where I_j is a set of intervals with size j in which the kth interval in I_j defines the interval between the kth symbol in the sequence and the symbol that follows I_j. They describe two algorithms for mining them (based on the Apriori and PrefixSpan algorithms) and constrain the mined time-intervals to a set possible time-intervals, as per the original algorithm by Chen et al. (2003). The algorithms are applied to two real datasets; a supermarket sales dataset and a dataset of stock market prices and volumes.

Another time-interval pattern type, called STP patterns (Álvarez et al. 2013), are defined by metric temporal constraint networks. A basic temporal constraint (x_1, i, x_2) is a time-interval constraint between two symbols x_1 and x_2 with time-interval i. These basic temporal constraints are combined together to form temporal constraint networks which are directed graphs, in which nodes represent symbols and edges represent basic temporal constraints. They describe an algorithm, ASTPminer, for mining frequent STP patterns that discovers basic temporal constraints by performing a cluster analysis of time differences between all pairs of nodes. Temporal constraint networks are then constructed with an Apriori-like strategy by iteratively combining basic temporal constraints if they meet the minimum support threshold. STPminer is applied to a real-world medical dataset of sleep apnea recordings, which are sets of time-stamped sequences.

From the **human/animal behaviour research community**, a different type of time-interval pattern was developed separately by Magnusson (2000). He introduces a hierarchical time-interval pattern called T-Patterns (see example in Fig. 1). T-Patterns define temporal dependencies by a critical interval (CI): $[d_1, d_2]$ between a pair of symbols (A, B) that exist if the occurrence B in time-interval $[t+d_1, t+d_2]$ is more likely to occur following an occurrence of A at time t than in a random interval of the same size. d_1 and d_2 are time offsets that take into account the allowable range of time differences. T-Patterns are defined as trees of CIs built in a bottom-up hierarchical fashion. Magnusson describes a mining algorithm that exhaustively searches all symbol pairs and for each pair, every possible CI is tested, from largest to smallest $(h, h - t, h - 2t, \ldots, 1)$ where h is the horizon and t is the time-step size, until a statistically significant CI is found. The dictionary of symbols grows, adding symbols to represent new CIs as they are discovered. The algorithm mines data from a sequence of time-stamped events, rather than transaction sets. A time-step increment t must be chosen for the CI mining algorithm (effectively rendering the data time-step discretised) as well as a maximum possible time-interval size h (i.e. the horizon). The statistical significance test requires knowledge of the total duration of data and the total number of events for each symbol. T-Pattern mining has been applied to a wide variety of human behaviour studies include schizophrenia, attention-deficit hyperactivity disorder, gender relation and courtship, communication and conversation, and sport (basketball, soccer swimming, martial arts). It has also been applied to animal behaviour analysis such as cognitive, foraging, courtship, anxiety, swimming behaviours. See Casarrubea et al. (2015) for a recent and comprehensive review of the application of T-Pattern mining.

Salah et al. (2010) extend Magnusson's mining algorithm, reducing the time complexity and the number of false positives returned. The exhaustive pair-wise search is carried out (as per the original algorithm) however only a single test is applied to find a CI for each pair (A, B), such computation time complexity is independent of the number of time-steps in $[1, h]$. The set of all time differences between A and B (within the horizon) are collected in a histogram and modelled as a 2-component Gaussian Mixture Model. Ideally, one component will be positioned at the sharp and localized peak of the CI, while the second, a broader, shallower peak will represent the remaining less statistically significant values. The width of the CI is derived from the standard deviation of the sharp peak. They test their method on a small synthetic dataset as well as real-world contact sensor data from a smart office environment.

Approaches from the data mining community generally mine time-interval patterns from transaction based databases and aim to extract the maximum number of patterns that meet a minimum support criteria. The hierarchical T-Pattern approach from human behaviour research, and its extension, mine from static datasets of time-stamped events (not segmented) and aim to extract one CI from each pair of symbols with the highest statistical significance. Both of these rely on the dataset being either segmented into transactions or static, with known data size (total duration and number of data points). Thus we propose, in the next section, a method for online mining of T-Patterns from an (unsegmented) event stream.

3   Mining T-Patterns from Event Streams

An event stream can be represented as a sequence of tuples $\{(x_1, t_1), (x_2, t_2), \ldots\}$, ordered by t, where x_i is an event, i.e. an instantaneous occurrence, of event type $S = \{A, B, \ldots\}$ at time $t_i \in \mathbb{R}$.

3.1   T-Patterns

We follow the definition of T-Patterns given by Magnusson (2000). T-Patterns are a hierarchical pattern made up of temporal relationships between pairs of event types (which we will also refer to as nodes). The temporal relationships, known as critical intervals (CI) define a time-interval $[t + d_1, t + d_2]$ between a pair of nodes $A, B \in S$. An occurrence of the CI takes place when an event of type A at time t is followed by an event of type B within $[t + d_1, t + d_2]$. To clarify, a CI refers to the time-interval relationship between two nodes and a T-Pattern refers to a hierarchical tree structure of CIs. However, each CI can also be described as a T-Pattern by considering the CI as the root of a T-Pattern tree that includes all of the CI's children.

For each CI there are two conditional probabilities, the forward probability $\frac{N_{AB}}{N_A}$ and the backward probability $\frac{N_{AB}}{N_B}$, where N_A and N_B are the number of occurrences of A and B respectively and N_{AB} is the number of occurrences of B that fall within the interval defined by the CI. In the original paper, these counts are taken from the static dataset, however from streaming data, this is not available. In this paper, once a CI pattern has been discovered, it maintains a count of all occurrences of N_A, N_B and N_{AB} from that moment onward.

In the original paper, $d_2 \geq d_1 \geq 0$, however in this paper we relax this constraint to $d_2 \geq d_1$ such that time-intervals that are negative or span zero are allowed. For example, in this paper, $[t - 0.5, t + 0.2]$ is valid, thus a CI does not necessarily imply the sequential order of the associated node pair (A, B). This is important to capture patterns of symbols that occur at similar times but not necessarily in a specific order, especially when dealing with noisy data.

3.2   Proposed Mining Algorithm

The mining algorithm can be summarised in Fig. 2, which runs periodically, according to a user specified frequency, each time resulting in one discovered CI.

Similarly to previous T-Pattern miners by Magnusson (2000) and Salah et al. (2010), when a new CI is discovered (i.e. added to the discovered pattern set) the dataset, which in our case is the state history, is updated to include occurrences of the two new event types that define the CI.

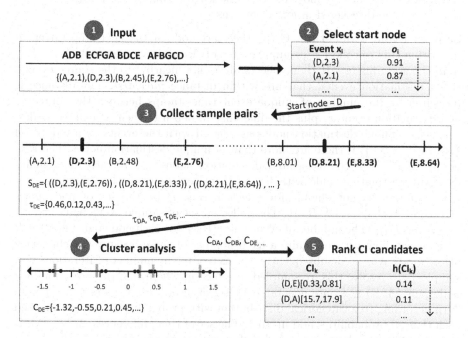

Fig. 2. Overview of proposed T-Pattern mining algorithm. First, select a start node. Next collect sets of sample event pairs from the state history. Denote the set of time difference between each pair of events as τ_{AB}. Perform cluster analysis on τ_{AB} to produce cluster sets C_{AB}. Convert cluster to CI candidates, rank them according to a heuristic score and the highest scoring CI_k is the result.

Start Node Selection. The algorithm begins by selecting a *start node* A by ranking the most recent n_s events from the state history, according to an event priority score. For the ith event x_i, event priority score $o_i = (\beta_d)^{d_i} * (\beta_s)^{s_i} * (\beta_p)^{p_i}$ where d_i is the depth of the associated node, s_i is the number of times the node has been the start node, p_i is number of parent CI's of the node and constants β_d, β_s and β_p, all in $[0, 1]$, determine the penalty weight of each these three components respectively. The node associated with the highest scoring event is chosen as the start node A. Note that this heuristic is employed further in the algorithm to prioritise discovery of patterns involving higher scoring events.

A random selection policy may suffice however we have found that this heuristic improves the outcome when tested in several small synthetic scenarios, especially where the task (or data distribution) changes significantly over time. The main motivation for this heuristic is to overcome challenges arising from the iterative constructive mining of a stream. The rationalisation is as follows. Sampling from recent events helps direct learning towards more recent observations in the stream. The d_i penalty prioritises shallow nodes to learn patterns involving nodes closer to input first then build deeper structures over time. s_i forces the priority to eventually move to deeper nodes over time and p_i prioritises nodes that have been discovered in fewer patterns.

Sample Event Collection. Given the start node A the next step (given in Algorithm 1) is to collect n_A sample events of A, called source events $(x_i \in E_A)$ from the state history. It then collects all events 'nearby' to the source events, called destination events, that are within distance max_dist. The distance is measured by the number of unique event time-stamps between the source x_i and destination event x_j plus the difference in depth of their associated nodes. The set of unique destination nodes associated with the events x_j are denoted $\{B_1, B_2, \ldots, B_m\}$. If there are more than m unique destination nodes then randomly select m of the associated destination nodes and discard the remainder. Any destination node with fewer sample events than the minimum support min_sup, is not eligible for selection. Let P be the set of node pairs $\{(A, B_1), (A, B_2), \ldots, (A, B_m)\}$. The collected set of sample event pairs (x_i, x_j) is denoted E_{AB}. The number of event pairs $|E_{AB}|$ for each node pair $(A, B) \in P$ is unknown, however we require a fixed number of these because the clustering algorithm that is used subsequently is sensitive to the number of iterations (i.e. samples). To resolve this, re-sampled sets RE_{AB}, where $|RE_{AB}| = n_{re}$ for each node pair $(A, B) \in P$, are generated by selecting n_{re} event pairs (x_i, x_j) from E_{AB} according to a categorical distribution with probability $o_i + o_j$, normalised such that all probabilities add up to 1. The set of time differences $\Delta t_{ij} \in \tau_{AB}$ is the time difference between x_i and x_j for each event pair $(x_i, x_j) \in RE_{AB}$.

Cluster Analysis. In this step a cluster analysis is performed on each of the time difference sets τ_{AB}, one set for each node pair (A, B). The output of this

---

Algorithm 1. Sample collection algorithm

Input: start node A

Output: Node pairs P, sample event pairs E_{AB} and time differences τ_{AB}

1 $S_A \leftarrow$ randomly select n_A events of start node A from state history

2 **for** *each event x_i in E_A* **do**

3 $\quad$ Collect all events x_j within distance max_dist of x_i and add event pair (x_i, x_j) to E_{AB} where B is the node of event x_j

4 $\{B_1, B_2, \ldots, B_m\} \leftarrow$ randomly select m unique destination nodes from event pairs $\in E_{AB}$, excluding destination nodes B for which $|E_{AB}| < min\_sup$

5 $P \leftarrow \{(A, B_1), (A, B_2), \ldots, (A, B_m)\}$

6 **for** *each node pair $(A, B) \in P$* **do**

7 $\quad$ $RE_{AB} \leftarrow$ select n_{re} event pairs (x_i, x_j) from E_{AB} according a categorical distribution with normalised probability $o_i + o_j$

8 $\quad$ $\tau_{AB} \leftarrow$ the set of time differences between x_i and x_j for all $(x_i, x_j) \in RE_{AB}$

---

step is a set of 1-dimensional cluster centres C_{AB} for each pair. Any clustering technique could theoretically be used here. We have chosen to use Self-Organising Map (SOM) algorithm (Kohonen 1982).

The SOM is a feed-forward neural network with a single computational layer of neurons that maps input vectors of arbitrary dimension onto a n-dimensional grid of output units. This mapping tries to preserve topological relations, i.e. patterns that are close in the input space will be mapped to units that are close in the output space, and vice-versa. Training of a SOM occurs by iteratively drawing samples from the input space and adjusting the weights of the "winning" neuron (i.e. the neuron that most closely resemble a sample input) such that it more closely matches the input. A key feature of the SOM algorithm is its neighbourhood function which also adjusts the weights of other neurons, that are nearby to the winning neuron in the output space, to be closer to the sample input. Each SOM neuron is a cluster center and thus a k-neuron SOM will perform a task similar to k-means clustering. The SOM and k-means algorithms are rigorously identical when the radius of the neighbourhood function in the SOM equals zero (de Bodt et al. 1997).

We choose the SOM algorithm for this clustering step because of its neighbourhood function (which draws non-winning cluster centres towards winning clusters), resulting in more clusters occupying the more dense areas of the sample space. This is desirable in our case because we are mostly interested in the more dense parts of the sample space, as these are expected to represent more consistent timing patterns.

Candidate Ranking. In the final step, candidate CIs are generated based on the cluster centres $c_k \in C_{AB}$ and ranked according to a heuristic score. The clusters are approximated by time-intervals whereby the boundaries are taken as the mid point between neighbouring cluster centres. The time-intervals are used to define candidate critical interval relationships (CIs) between the node pair (A, B). Specifically, we derive $|C_{AB}| - 2$ critical intervals, where the kth interval $CI_k = [d_1, d_2]$, $d_1 = (c_k + c_{k+1})/2$ and $d_2 = (c_{k+1} + c_{k+2})/2$. The candidate CIs are ranked according to heuristic score

$$h(CI_k) = \frac{\sum_{(x_i, x_j) \in E_{AB}^k} h(x_i, x_j)}{|E_{AB}^k|} (\beta_{sd})^{\sigma(\tau_{AB}^k)}$$

where

- E_{AB}^k is the set of event pairs $(x_i, x_j) \in E_{AB}$ for which time difference Δt_{ij} falls within time-interval k
- $h(x_i, x_j) = \frac{\Delta P(x_j)(o_i + o_j)}{|E_{AB}^{ki}|^2}$ is a heuristic score for event pair (x_i, x_j)
- $|E_{AB}^{ki}|$ is the number of event pairs in E_{AB}^k that x_i is a member of
- Probability improvement: $\Delta P(x_j) = P_k - \hat{P}(x_j)$, if $P_k - \hat{P}(x_j) > min\_imp$, otherwise it is 0. Constant min_imp defines the min threshold for probability improvement, P_k is the forward probability of CI_k and $\hat{P}(x_j)$ is the maximum probability given by any parent CI of event x_j (either forward or backward)

– $\sigma(\tau_{AB}^k)$ is the standard deviation of values in τ_{AB}^k, constant $\beta_{sd} \in [0,1]$ determines the penalty weight for standard deviation of time difference values.

This heuristic score is rationalised as follows. Candidates with a higher density of samples are prioritised with $(\beta_{sd})^{\sigma(\tau_{AB}^k)}$, as these are expected to represent more consistent timing patterns. Candidates that frequently have more than one destination event x_j (within the interval) for each source event x_i are penalised via $|E_{AB}^{ki}|^2$. The score limits redundant patterns with $\Delta P(x_j)$ because destination events x_j that can be predicted by already discovered patterns with higher probability $\hat{P}(x_j)$, don't improve $h(CI_k)$.

The final output of the mining algorithm is the single highest ranking candidate CI, if $h(CI_k) > 0$, which is added to the discovered pattern set.

4   Evaluation and Results

As with any new architecture, it is typical to do the proof-of-concept with simulated examples, which not only provides clear abstraction but also allows for incremental and controlled testings. We use a benchmark synthetic dataset by Fellous et al. (2004), modified to simulate the conditions of online learning from a stream. We do not aim to demonstrate an improved performance over existing algorithms (because they are not directly comparable), only to demonstrate the capability of proposed algorithm to mine T-Patterns, online from event streams. For the experiments presented in this paper, the proposed T-Pattern mining algorithm runs periodically, between periods of observing the input stream of events. The state history is also culled periodically if it exceeds a fixed limit.

4.1   Fuzzy V-Measure for Evaluation

To evaluate the proposed algorithm, we consider each discovered T-Pattern as a cluster, c_i, that belongs to the set of discovered T-Patterns C. We then compare C to the ground truth, a set of known classes (G), to measure the success of this unsupervised learning task. In this regard, the discovered pattern set is interpreted as a soft clustering, producing a fuzzy cluster assignment for each item (where the item may be assigned a non-zero value for more than one cluster and their sum may not be equal to 1). The proposed incremental algorithm adds to the discovered pattern set over time and as a result, $|C|$ varies and is not necessarily equal to $|G|$. Due to the constructive hierarchical nature of the proposed mining approach, many more clusters than the number of classes tend to be produced (e.g. a desirable cluster may require many child clusters that do not match a known class), however we only require a match of one cluster for each class. Note that the proposed algorithm does not know the number of known classes $|G|$.

Previous results on this dataset use different performance measures. Results from Fellous et al. (2004), Paiva (2008), where $|C|$ is fixed in advance to equal $|G|$, report the percentage of trials correctly attributed to their original classes.

Results from Humphries (2011), where the cluster analysis is performed as community detection and the number of found groups $|A|$ is not necessarily equal to the real number of groups $|B|$, report a normalised mutual information measure between group structures A and B. These methods are not appropriate in our case; the former because it assumes $|C| = |G|$ and the latter because clusterings with $|C| \neq |G|$ are punished.

We use the Fuzzy V-Measure with dissimilarity (Utt et al. 2014), which extends the V-measure (Rosenberg and Hirschberg 2007), to measure the similarity between the learned clusters C and the known class labelling G, in which both C and G can be fuzzy and a perfect clustering score (of 1) can be achieved even when $|C| > |G|$. The measure is a weighted harmonic mean of two complementary properties—*homogeneity* and *completeness*, both of which are entropy-based. Homogeneity, $hom(C) = 1 - \frac{H(C|G)}{H(C,G)}$, if $H(C,G) > 0$, otherwise $hom(C) = 1$, is maximum when each cluster contains only members of a single class. H is the standard measure of entropy, $H(C|G)$ is the conditional entropy of C given G and $H(C,G)$, is the joint entropy and is used for normalization. Completeness, $com(C) = 1 - \frac{H(G|C)}{H(G,C)}$, if $H(G,C) > 0$, otherwise $com(C) = 1$, is maximum when all members of a given class are assigned to the same cluster. The entropy values are calculated with the joint probability of a cluster c_i and known class g_j, $\hat{p}(c_i, g_j) = \frac{\alpha_{ij}}{M}$ where M is the total mass of the cluster c_i and the mutual evidence table α_{ij} is constructed as follows. A one-to-one mapping is forced between C and G by choosing the closest matching cluster c_i to each class g_j based on a score that incorporates *similarity* and *dissimilarity*: $sim_{ij} - dis_{ij}$, where similarity $sim_{ij} = \mu(c_i \cap g_j)$ is the shared mass of c_i and g_j, dissimilarity $dis_{ij} = \mu(c_i) + \mu(g_j) - 2\mu(c_i \cap g_j)$ is the missing and remaining elements of the combination and mass function μ determines the mass of each data point as 1 divided by the number of classes it belongs to. The mutual evidence table α_{ij} is constructed by first setting all values to 0 except for the best mapping entries, for which we set $\alpha_{ij} = sim_{ij}$. The error mass (dis_{ij}) for each of the best mapped entries is then distributed evenly over the non-zero entries in each row.

4.2 Experiment with JN2004 Synthetic Data

We use synthetic data[1] created by Fellous et al. (2004) which is designed to test clustering of segments of neural spike trains (i.e. time-stamped event sequences) with controlled amounts of noise. The data consists of sets of synthetic event segments where each segment (or item) belongs to one of the known classes G. A class is defined by E event times chosen uniformly at random within a 1 s interval (e.g. a class might be defined by $E = 5$ events (at times 0.07, 0.23, 0.31, 0.62, 0.84)). Three parameters control noise added to each item: each event time is jittered by an amount drawn from a Gaussian distribution of mean 0 and standard deviation J ms; X extra events (that don't belong to the class) are added to the item at times sampled uniformly from the same 1 s interval; and each of the class events have an M probability of not being included in the item.

---

[1] Available from http://www.snl.salk.edu/~fellous/data/JN2004data/data.html.

Various levels of noise are defined by a combination of J and X values taken from Humphries (2011) where $J \in \{0, 1, 3, 5, 10\}$ and $X \in \{0, 2, 3, 4, 8\}$. For each increment in noise level $I \in [0, 1, \dots, 4]$, the values of J and X are simultaneously incremented (e.g. at noise level of 3 $J = 5$ and $X = 4$). See example in Fig. 3.

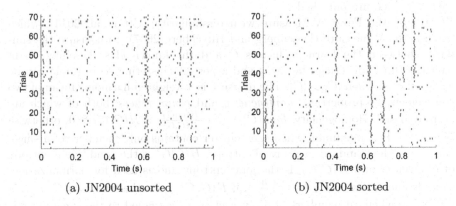

(a) JN2004 unsorted   (b) JN2004 sorted

Fig. 3. Raster plots of example synthetic event sequences and resulting clustering. (a) The original dataset with 70 items (trials), 35 for each of $|G| = 2$ clusters with noise level I = 3 ($J = 3$ ms and $X = 3$ extra events) and $M = 0.15$ probability of missing events. (b) Correct grouping of the items into two clusters

The experiment is run as follows. The proposed system observes the input stream until 50 'rounds' of items have been presented, whereby in each round one item from each known class in G is presented in random order with random time delays between them. The division into rounds is done for evaluation purposes only. Note that the learning system has no knowledge of the round or item boundaries nor the number of rounds, items or known clusters. After the first 50 rounds of items are presented (allowing the state history to accumulate) the input stream continues in the same manner and one instance of the mining algorithm is run between every round. This configuration sets up the scenario of online learning from a stream, however the results are not sensitive these parameters except for having enough data in the state history to draw samples from. The mining continues in this fashion for 70 rounds.

Figure 4(a) shows the Fuzzy V-Measure score ($\in [0, 1]$) after each round of mining for one experiment with parameters $M = 0, E = 5, |G| = 5$ known classes and noise level $I = 0$. A total of 49 T-Patterns were learned after 70 round of learning (potential maximum of 1 per round but in some rounds the highest ranking candidate did not meet the minimum score). After 34 rounds of mining, a Fuzzy V-Measure score of 1 is reached, demonstrating that five of the 49 T-Patterns learned by the proposed mining algorithm, online from the event stream, corresponded exactly to the five known classes. Each of these five hierarchical T-Patterns, consisted of either 6 or 7 individual CI relationships.

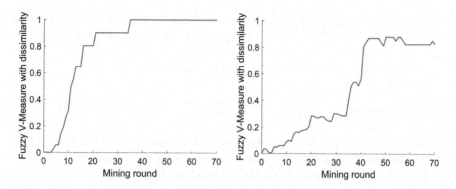

Fig. 4. The performance of the algorithm over 70 rounds of mining, measured by Fuzzy V-Measure with dissimilarity. (a) Experiment with parameters $M = 0, E = 5, |G| = 5$ classes and no noise $I = 0$. (b) Experiment with parameters $M = 0, E = 5, |G| = 3$ classes and noise level $I = 1$

In a second experiment with $|G| = 3$ and noise level $I = 1$ the results in Fig. 4(b) show that T-Patterns were learned that correspond to the known classes but to a lesser degree. In the final round, the Fuzzy V-Measure analysis showed that for 2 of the 3 known classes, there were 2 learned T-Patterns that matched them with a perfect score. The best matching cluster for the remaining known class attributed around 83% of this class's item mass to the correct class and the remaining items to the other 2 classes.

To measure the performance of the mining algorithm with different numbers of known classes and for various noise levels, we ran 20 experiments for each combination of $|G| = [2, 3, 5]$ and $I = [0..4]$. The Fuzzy V-Measure score from the

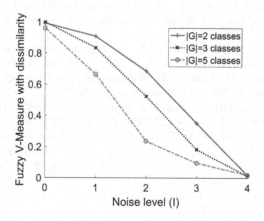

Fig. 5. Clustering results using parameters $|G| = [2, 3, 5], M = 0, E = 5$ over various noise levels $I = [0..4]$. The experiment was run 20 times for each combination of parameters and results were averaged.

final round only, is taken and averaged over 20 experiments for each combination of parameter settings. The results in Fig. 5 show that the proposed approach is capable of reaching a perfect or near perfect score with no noise ($I = 0$) and exhibits graceful degradation as more noise is added ($I = [1..4]$).

5   Conclusion

In this paper we introduced an algorithm for mining T-Patterns, online from unsegmented event streams. The key ideas are: to perform incomplete searches by clustering data sampled from the stream and to perform iterative mining of CIs to discover hierarchical T-Pattern structures. Experiments were carried out on a benchmark synthetic dataset that demonstrated the algorithms ability to discover appropriate T-Patterns by comparing them to a known gold standard. The experiments also demonstrated that the algorithm exhibits graceful degradation with respect to added noise. In future analysis work we aim to evaluate the proposed approach on real world data sets and analyse the time and space complexity. We will also aim to develop a pruning policy to limit the total size of the model (i.e. discovered pattern set) when running indefinitely and an algorithm for mining temporal association rules from event streams.

References

Agrawal, R., Srikant, R.: Mining sequential patterns. In: Proceedings of the Eleventh International Conference on Data Engineering. IEEE (1995)

Álvarez, M.R., FéLix, P., CariñEna, P.: Discovering metric temporal constraint networks on temporal databases. Artif. Intell. Med. **58**(3), 139–154 (2013)

Casarrubea, M., Jonsson, G., Faulisi, F., Sorbera, F., Di Giovanni, G., Benigno, A., Crescimanno, G., Magnusson, M.: T-pattern analysis for the study of temporal structure of animal and human behavior: a comprehensive review. J. Neurosci. Methods **239**, 34–46 (2015)

Chen, Y.-L., Chiang, M.-C., Ko, M.-T.: Discovering time-interval sequential patterns in sequence databases. Expert Syst. Appl. **25**(3), 343–354 (2003)

de Bodt, E., Verleysen, M., Cottrell, M.: Kohonen maps versus vector quantization for data analysis. In: ESANN, vol. 97, pp. 211–218 (1997)

Fellous, J.-M., Tiesinga, P.H., Thomas, P.J., Sejnowski, T.J.: Discovering spike patterns in neuronal responses. J. Neurosci. **24**(12), 2989–3001 (2004)

Hu, Y.-H., Huang, T.C.-K., Yang, H.-R., Chen, Y.-L.: On mining multi-time-interval sequential patterns. Data Knowl. Eng. **68**(10), 1112–1127 (2009)

Humphries, M.D.: Spike-train communities: finding groups of similar spike trains. J. Neurosci. **31**(6), 2321–2336 (2011)

Kohonen, T.: Self-organized formation of topologically correct feature maps. Biol. Cybern. **43**(1), 59–69 (1982)

Magnusson, M.S.: Discovering hidden time patterns in behavior: T-patterns and their detection. Behav. Res. Methods Instrum. Comput. **32**, 93–110 (2000)

Paiva, A.R.: Reproducing kernel Hilbert spaces for point processes, with applications to neural activity analysis. Ph.D. thesis, University of Florida (2008)

Rosenberg, A., Hirschberg, J.: V-measure: a conditional entropy-based external cluster evaluation measure. In: EMNLP-CoNLL, vol. 7, pp. 410–420 (2007)

Salah, A.A., Pauwels, E., Tavenard, R., Gevers, T.: T-patterns revisited: mining for temporal patterns in sensor data. Sensors 10(8), 7496–7513 (2010)

Utt, J., Springorum, S., Köper, M., Im Walde, S.S.: Fuzzy V-measure-an evaluation method for cluster analyses of ambiguous data. In: LREC (2014)

Outlier Detection and Applications

Detection of Outlier Behaviour Amongst Health/Medical Providers Servicing TAC Clients

Musa Mammadov[1]([⊠]), Rob Muspratt[2], and Julien Ugon[1]

[1] School of Science and Technology, Federation University,
Ballarat, VIC 3350, Australia
{m.mammadov,j.ugon}@federation.edu.au
[2] Transport Accident Commission, Geelong, VIC 3220, Australia
rob_muspratt@tac.vic.gov.au

Abstract. Within the landscape of Personal Injury Compensation, building of Decision Support Tools that can be used at different stages of a client's journey, from accident to rehabilitation, and which have various targets is important. The challenge considered in this paper is concerned with finding outliers amongst Health/Medical Providers (providers) servicing Transport Accident Commission (TAC) clients. Previous analysis by the TAC in this domain has relied upon data aggregation and clustering techniques and has proven to be restrictive in terms of providing easily interpretable and targeted results. In particular, the focus of this study is to identify outlying behaviours amongst providers rather than individual exceptional cases. We propose a new approach that enables identification of outliers on the basis of user defined characteristics.

Keywords: Personal Injury · Compensation
Health/medical provider · Outlier behaviour · Input-output space
Combination targeting

1 Introduction

The TAC is a Victorian Government-owned organisation whose key functions are funding treatment and support services for people injured in transport accidents, promoting road safety in Victoria and improving the state's trauma system. The TAC is a "no-fault" scheme. This means that benefits and support services will be paid by the TAC on behalf of an injured person (client/claim) irrespective of who caused the accident.

The TAC is committed to protecting the funds and other assets placed under its control for the benefit of the people of Victoria and hence uses analytical methods/techniques to monitor the payment of benefits and services to clients and providers. It is the role of the TAC Forensics Group and Provider Review Team to conduct this monitoring and subsequent review of outlier behaviour,

© Springer Nature Singapore Pte Ltd. 2018
Y. L. Boo et al. (Eds.): AusDM 2017, CCIS 845, pp. 161–172, 2018.
https://doi.org/10.1007/978-981-13-0292-3_10

with a particular focus on health/medical provider billing. Further details regarding this program of work can be found at the TAC website under *Billing Review Program - Information for Medical Practitioners* (TAC 2016).

Prior methods applied internally by the TAC for monitoring provider billing and behaviours have focused on data aggregation and comparison at the provider level. Techniques employed have included rules based exceptions, weighted rankings, discipline measure comparison or set analysis, clustering and exploration via Self-Organising Maps (SOMs). Aggregation at the provider level has often resulted in a loss of outlier "sensitivity", particularly with regard to individual provider-claim combinations. Indeed, premature aggregation can eliminate all distinction between outlying behaviour and underlying exceptional cases.

A collaborative effort between the TAC Forensics Provider Review Team and Federation University has enabled the development of the method presented below and which helps alleviate this limitation of previous monitoring techniques.

2   Outlier Detection Methodologies

An outlier is an observation (or measurement) that is different with respect to the other values contained in a given dataset (Barnett et al. 1994; Cateni et al. 2008; Chen et al. 2002; Hawkins 1980; Hodge and Austin 2004; Ramaswamy et al. 2000). The problem of detecting outliers has important applications in many areas, including credit card fraud, network robustness analysis, network intrusion detection, financial applications and marketing (Cateni et al. 2008; Hodge and Austin 2004).

There are a large number of publications and methods developed for outlier detection. Distribution-based, depth-based, clustering, distance-based or density-based methods deal with the geometrical distribution of data points. On the other hand, many artificial intelligence methods can be used to find outliers by employing different techniques, like Neural Networks, Support Vector Machines, Principal Component Analysis, Self-Organising Maps, Least Squares and Linear Regression.

2.1   Industrial Applications

Despite a large number of methods having been developed for outlier detection, their successful implementation hugely depends on the application area. One of the most important steps in these implementations is data generation; that is, collecting the most relevant and useful information that may help in finding outliers. The choice of a suitable method (say out of those mentioned above) could be considered as a next step. However, the major task of finding outliers is still far from being solved even after data collection and selecting the method to be applied. This is the main aspect which we aim to explore in this paper, with a particular application to outliers amongst health/medical providers.

It would be expected that any outlier detection scheme will encounter several serious concerns and/or difficulties. We mention here a few, quite common ones, that will be in our focus throughout the paper:

C1: feature/variable selection;

C2: deriving new variables or dimensionality reduction (e.g. by Principal Component Analysis);

C3: an appropriate rescaling of variables selected for the analysis;

C4: the choice of appropriate distance measures.

Clearly, any variations in these considerations may alter significantly the list of outliers obtained by a particular algorithm/method. In many cases the parameter tuning (e.g. an appropriate number of clusters in clustering-based methods) is an extremely difficult task.

On the other hand, considering different sets of parameter settings and generating corresponding sets of outliers may not be desirable due to the presumably large number of outliers that would not be possible to process further (limited resources for review activities).

Moreover, even after dealing with these issues, the outliers obtained need to be analysed and the outlying reasons need to be understood in order to guide/help further investigations.

3   A New Approach Based on Input-Output Space Representation

Taking into account the above mentioned concerns, in this paper we introduce a new approach that deals with these considerations with greater clarity. This approach compares providers in terms of their actions or behaviours over similar cases/claims. Simply put, it utilises the following scheme:

$$Input \rightarrow Provider \rightarrow Output$$

Input is assumed to combine a set of features that could be used to define "similar clients". For example, age, gender, postcode, injury types, etc.

Output is assumed to be a set of responses/actions by a particular *Provider*. For example this may include service types, service levels, service intensities, billed amount, type of billing, etc. Output here should not be confused with outcome. Depending on the focus of the study, outcome may be considered an output, but some output, such as cost or type of service, cannot be considered an outcome.

In this approach, the available data and corresponding variables are considered in terms of their suitability to describe "what is given" (that does not depend on providers) and "what is the output" that entirely depends on provider's choice, action or performance. In this way, we have the following advantages:

1. It assists us in selecting variables, as well as deriving new variables that can describe the input and the output factors better (concerns C1, C2).
2. It makes it possible to consider input and output variables (spaces) separately and to deal with the concerns C3, C4 (that is, scaling and the choice of distance functions) separately and more efficiently.

These advantages will be discussed in detail below and implemented by considering 12 months of TAC provider billing data (Sample Provider Group) for a specific provider discipline. In addition we note that the above scheme can be considered as a user specified "scenario" where, for example, the outlying criteria (e.g. *Output = Amount Paid*) can be predefined depending on user's interests.

3.1   Algorithm

We denote the data for our study by $D = \{(p_k, c_{kl}; x_{kl}, y_{kl}) : k = 1, \cdots, N, l = 1, \cdots, n(p_k)\}$ where

- N is the number of providers;
- p_k is the provider IDs;
- $n(p_k)$ is the number of claims by provider p_k;
- c_{kl} is the claim ID(s) by provider p_k;
- x_{kl} and y_{kl} are the input and output variables.

We will use different distance functions for input (x) and output (y) variables that will be denoted by d^x and d^y, respectively. We also set n^{top} as the number of close claims or the size of the neighbourhood.

Step 1: Given a provider-claim combination (p, c), this step introduces two criteria measuring the degree of abnormality. Consider an arbitrary data point $(p^0, c^0; x^0, y^0)$ in D.

- Calculate the distance $d^x(x^0, x)$ from all data points $(p, c; x, y)$ and select the closest n^{top} points, the neighbourhood, which will be denoted by N^0.
- Calculate the average values AvS^1 and AvS^2 of distances $d^x(x^0, y)$ over all data points N^0 and $D\backslash N^0$, respectively.
- Calculate the following two outlying values:
  - $OM^1(p^0, c^0) = AvS^1$; and
  - $OM^2(p^0, c^0) = AvS^1 - AvS^2$.

The first value defines outliers in terms of "local" neighbourhood, that is, the divergence with respect to the closest n^{top} claims. In contrast, the second value is "global"; it considers also the divergence from the other (distinct) claims. Then, a provider-claim combination (p^0, c^0) is considered an outlier, if the corresponding output variable y^0 is "closer" to the outputs of the set $D\backslash N^0$ rather than N^0.

Step 2: Arrange outlying values $OM^1(p_k, c_{kl})$ and $OM^2(p_k, c_{kl})$ calculated in Step 1 in descending order. Let provider-claim combinations (p_k, c_{kl}) be arranged in the form (p_k^1, c_k^1) and (p_k^2, c_k^2), $k = 1, \cdots, |D|$ (here $|D|$ is the size of D), such that

$$OM^1(p_1^1, c_1^1) \geq \cdots, OM^1(p_{|D|}^1, c_{|D|}^1) \tag{1}$$

and

$$OM^2(p_1^2, c_1^2) \geq \cdots, OM^2(p_{|D|}^2, c_{|D|}^2). \tag{2}$$

Step 3 (Provider level): At this level, for each outlying value OM^m ($m \in$ $\{1, 2\}$), we divide the provider-claim combinations into 5 different subsets: L^m_{top1}, L^m_{top2}, L^m_{mid}, L^m_{bott2} and L^m_{bott1} according to the order in (1) and (2); where $top1, top2$ and $bott2, bott1$ correspond to the top N_1, N_2 and bottom N_2, N_1 cases, L^m_{mid} combines all the cases that are not in $L^m_{top1} \cup L^m_{top2}$ and the corresponding outlying value OM^m is greater than the average over all data points. Accordingly, given provider p, we denote by $n^m_p = (n^m_{p1}, \cdots, n^m_{p5})$ the number of occurrences in these subsets.

The main idea in considering the top two subsets is as follows. Practice shows that the highest ranked outliers usually have solid reasons for their large outlying values OM^m (as exceptional cases). Then the most interesting outliers could be expected to be in L^m_{top2} rather than in L^m_{top1}. Taking this into account, we can apply weights $w = (w_1, \cdots, w_5)$ for the lists L^m_{top1}, L^m_{top2}, L^m_{mid}, L^m_{bott2} and L^m_{bott1}, respectively. Following the explanation above, the weight w_2 could be expected to be larger than w_1.

After introducing weights w, we can quite easily summarise the outlying values for a given provider p in the linear form:

$$OM^m(p) = w_1 n^m_{p1} + \cdots + w_5 n^m_{p5}, \ m = 1, 2.$$

Finally, for each m, one can flag outliers according to the values $OM^m(p)$. We recall that $m = 1$ and $m = 2$, in some sense, correspond to "local" and "global" outliers, respectively.

By setting different weights w, outliers can be segmented with different purposes in mind. For example, if the output variable y describes the "number of services" then the weight vector $w = (1, 5, 0, 0, 0)$ would define outliers in terms of "over-servicing" and/or "under-servicing". Observations in L^m_{bott2} and L^m_{bott1} can therefore be considered the most "normal" cases and not a focus for this application.

3.2 Advantages of the Proposed Approach

The advantages of this approach are discussed below by considering the concerns C1–C4 mentioned above.

Feature/Variable Selection and Deriving New Variables. The list of available and useful feature/variables includes:

- About Provider: *postcode, qualification codes, tenure, previous investigation/audited flag, favoured service type/item, client base complexity rating, denied services, growth measures;*
- About Claim: *postcode, gender, age, accident date, list of injuries* (including the *main injury*);
- About services (for each visit): *service date, service type, service location, account received date, payment date, payment type, billed amount, paid amount.*

Quite useful derived-variables that could be considered include (for a particular Provider-Claim and given time interval):

- *time from accident = service date − accident date;*
- *time to receipt = account received date − service date;*
- *duration = last service date − first service date;*
- *service intensity = number of services/duration.*

Input Variables. These variables are used to select similar claims among all the data points. After analysing the variables from the Sample Provider Group it was observed that the following are key features to be used for this aim:

I1: List of Injuries. The injuries of a client presumably define the type of services that could be expected. They are also useful to estimate the duration a particular type of service may be required for (years or months). Many TAC datasets use only the *main injury* among all types of injuries. We use the list of all 20 injuries (details are in Table 2, Appendix) applied by TAC. In this way all the secondary injuries are taken into account. The resulting *Injury Vector* $(I_1, \cdots, I_{20})$ is a binary vector of length 20 with a minimum of 1,800+ combinations for the Sample Provider Group in question.

Injuries are amalgamated from the following sources: *Snomed* codes entered into the TAC Claims Management System at claim lodgement/acceptance; ICD10 codes captured when hospital accounts are processed by the TAC; Internal Client Service Profiles for severely injured clients; historical ICPC Injury Codes archived from the previous TAC Claims Management System (prior to 2008).

Note: The idea behind using all injuries is thus. It is quite possible to experience a full recovery from the *main injury;* subsequently the reason for attending a medical provider might be a "secondary" injury that exhibits a longer recovery time (e.g. *Degloving, Burns,* etc.).

I2: Age. Clearly, *Age* is an important feature of any claim. Recovery rate and medical conditions may heavily depend on it.

I3: Time from Accident. In many instances, *Time from Accident* plays an important role in a client's journey and may be a good indicator of the type of services that one could expect to fund.

Output Variables. Given our definition above these variables should describe a provider's action given a particular claim. This means that all the variables related to services rendered could be considered for this aim.

Despite a large number of variables listed above, one can easily observe that only a few of them should be selected as key factors that differentiate provider's actions. They are:

O1: Services. We refer to the *Medicare Benefits Schedule* (MBS) (2016) that lists the 274 different service types evident in Sample Provider Group data. Only a few of these are frequently used; for example, the top 20 services were observed to cover 95.4% of the data. Services used infrequently can be excluded once it is confirmed that they are not in scope for review.

There are similar types of services/clusters that are differentiated by the service duration-time. The *MBS* (2016) defines 4 different levels A, B, C and D for the services provided by General Practitioners and 5 different levels A, B, C, D and E for the services provided by Psychiatrists. In many instances, not the type of service (like *Surgery Consultation, Home Visit*, etc.), but the level of service depends on the choice made by a provider or clinic; thus they can better describe a provider/clinic's actions.

There are usually many visits/services related to a provider-claim combination. These services may involve different types or levels. We use a vector of services $s = (s_1, \cdots, s_L)$ for each provider-claim combination. Here L is the number of service types or levels, and s_i is the number of services of type/level $i \in \{1, \cdots, L\}$.

O2: Service Intensity/Frequency. This is another useful feature that can be applied to differentiate providers. Given that we are considering one years of billing data it is quite natural to define the service intensity as an average number of visits per month. This feature is complementary to the service levels and describes providers from a different point of view. For example, a *high* intensity and *low* intensity could be classified as *over serviced* and *under serviced*, respectively.

O3: Amount Requested/Paid. The amount requested by a provider, in some cases, is different from the amount actually paid. Taking this into account, we did not use the difference between the amount requested and paid to describe abnormalities since, in the Sample Provider Group data, rates are generally in line with that of the MBS Guide. Instead, we used the amount paid (by TAC) as a more accurate amount for a given service.

Similar to the service types, we aggregated the amounts related to a particular provider-claim combination; that is, we used the *average amount paid.*

The *average amount paid* is a complementary feature to service levels. Although one can expect a strong correlation between these outputs, the amount paid (as a scalar number in dollars) can be added together to derive an average amount per a provider-claim combination. Clearly, it is also complementary to the *service intensities* that measure the average number of services per month.

It is asserted that these three measures describe all the possible "aspects" of provided services where a provider's selection/action is concerned. Taking this into account, no further variables are considered appropriate for *Output.* It is also worth noting that variable selection in all cases was conducted to minimise the complexity of processing/interpretation and maximise relevance of the output. Additional features could very well be included in the future, such as geographical location (derived from *postcode*), if they are considered relevant to the task and do not compromise the statistical significance of population sub-groups.

Rescaling and the Choice of Appropriate Distance Measures. This section addresses the concerns C3 (rescaling) and C4 (distance measures) considered in Sect. 2.1. We consider *Input* and *Output* variables separately and show that the approach suggested in this paper helps us greatly when dealing with these concerns. In fact, it turns out that for the *Output* variables a choice of rescaling is not required.

(1) Input variables: *Age, Time from Accident* and *Injury Vector* (list of 20 injuries).

Rescaling. Injury Vector $(I_1, \ldots, I_{20})$ is constructed with the severity levels in descending order of severity. Therefore, in terms of defining close claims, it would be useful to differentiate the severity of injury types. This can be performed by applying weights to injury types such that the higher level injuries are associated with higher weights. We used the following (exponential) formula (see Table 2, Appendix):

$$I'_K = 100 I_k / k^2, \quad k = 1, \cdots, 20 \tag{3}$$

to generate a continuous weighted injury vector $(I'_1, \cdots, I'_{20})$.

Time from Accident corresponding to different services for a particular provider-claim combination, are averaged to derive one scalar value - the *Average Time from Accident* in order to describe the claim's age.

Finally, the *Age* and the *Average Time from Accident* are rescaled so that they have the same range [0, 5]. In this case, 1 unit (weight for injury level 10, see (3)) corresponds to approximately 20 years (in *Age*) and 8 years (in *Time from Accident*).

Distance. The distance function is used to define similar claims; that is, to select a close neighbourhood of a given claim. The Euclidean distance would naturally be the most suitable measure for this aim and hence choice of distance measure is not a major concern in this case.

(2) Output variables: *Service Types* or *Service Levels*; *Service Intensities* and *Average Amount* per provider-claim combination.

Service Intensities and *Average Amount* are scalar variables. Accordingly, there is no need to choose any particular distance function or to rescale the data.

When using the variable *Service Types* or *Service Levels* the major interest is in the proportion between these components. This means that there is no need to rescale the data. Moreover, the Cosine distance is the most suitable measure that is often used in similar problems where the proportion of components in a vector is the major focus (like in text categorization). We will use the Cosine distance that is defined as follows: given two service vectors s^1 and s^2,

$$D^c(s^1, s^2) = 1 - \cos(s^1, s^2).$$

Therefore, in terms of concerns C3 and C4, an appropriate scaling of *Input* variables and distance calculation of *Output* variables are the only serious issues needing to be tackled.

4   Numerical Results

Input data derived from Sample Provider Group billing contained 98,190 observations or services covering a single 12 month period. This data is comprised of 6,045 unique providers, 14,922 unique claims and 22,120 provider/claim combinations.

Run1: In the first run the top 30 most frequent services for service vector $s = (s_1, \cdots, s_L)$ were used. Results obtained revealed several clearly prominent outlier providers. Closer investigation showed that they became outliers due to the specific type(s) of services they provided; including writing reports and impairment examinations. There were 23 such services.

Run2: Since the main provider outliers related to the 23 services mentioned above were not in scope of review, all data points involving these services were removed. The number of data points was reduced by 1.8% to 96,417 observations. Running the same algorithm again obtained several very clear outliers. Further investigation showed that many of these outlier providers were related to quite special types of services. They included *Registrar at Hospital, Rehabilitation Specialist* and *Emergency Physician*. 11 providers were identified in these categories.

Run3: In the final run all the data relating to the 11 (special) providers mentioned above were also removed. Moreover, instead of considering the top 30 most frequent services separately, we selected all services that specify service duration only. In this way, we used corresponding service levels covering approximately 91% of data (89,556 out of 98,190 observations). Two weighting vectors $w = (10, 20, 3, 0, 0)$ and $w = (1, 2, 0, 0, 0)$ were also considered for the final ranking with output for the former shown in Table 1 below.

Table 1. Sample provider group - outliers

Provider ID	Ob1 Service Levels	Ob2 Servs/Month	Ob3 Avg-amount	Ob1 Service Levels	Ob2 Servs/Month	Ob3 Avg-amount	Score	SL Top500	SL Top[501	SL Rest	SL Top%	Int Top500	Int Top[501	Int Rest	Int Top%	Avg Top500	Avg Top[501	Avg Rest	Avg Top%
426	+	+	+	+	+	+	6	1	3	8	33.3%	0	4	8	33.3%	1	1	10	16.7%
1221	+	+	+	+	+	+	6	0	5	71	6.6%		3	69	9.2%	2	3	71	6.6%
1272	+	+	+	+	+	+	6	0		48	25.0%	1	3	60	6.3%	1		53	17.2%
2118	+	+	+	+	+	+	6	0	6	31	16.2%	0	3	34	8.1%	0	5	32	13.5%
837	+	+	+	+	+	+	6	0	1	23	4.2%	2	4	18	25.0%	0	3	21	12.5%
950	+	+	+	+	+	+	6	2	2	18	18.2%	1	1	20	9.1%	0	3	19	13.6%
1076	+	+	+	+	+	+	6	2	3	20	20.0%	2	4	19	24.0%	3	2	20	20.0%
1195	+	+	+	+	+	+	6	0	2	38	5.0%	0		28	30.0%	0	0	40	0.0%
1276	+	+	+	+	+	+	6	0	0	32	0.0%	0	1	31	3.1%	0	1	31	3.1%
1729	+	+	+	+	+	+	6	1	4	29	14.7%	0	1	33	2.9%	0	3	31	8.8%
1874	+	+	+	+	+	+	6	0	5	52	8.8%	1	5	51	10.5%	0	3	54	5.3%
1963	+	+	+			+	4	0	5	16	23.8%	1	3	17	19.0%	1	3	17	19.0%
2077		+		+	+	+	4	2	8	57	14.9%	0	2	65	3.0%	0		60	10.4%
3348	+		+	+		+	4	0	4	5	44.4%	0	1	8	11.1%	1	3	5	44.4%
4763	+	+	+			+	4	0	0	17	0.0%	1	3	13	23.5%	0	2	15	11.8%
107	+	+	+			+	4	0	1	13	7.1%	0	0	14	0.0%	0	1	13	7.1%
538	+	+	+			+	4	1	2	22	12.0%	0	3	22	12.0%	0	4	21	16.0%
638	+	+	+	+		+	4	2	3	7	46.2%	0	0	13	0.0%	0	2	11	15.4%
684	+	+		+			3	0	3	10	23.1%	1	2	9	30.8%	1	4	8	38.5%
755	+	+		+			3	1	0	8	11.1%	0	1	8	11.1%	0	1	8	11.1%
1089	+	+	+				3	0	3	6	33.3%	0	0	9	0.0%	1	1	7	22.2%
1134	+	+	+				3	0	1	33	2.9%	0	5	29	14.7%	0	2	32	5.9%
1319			+	+	+	+	3	2	2	31	11.4%	0	3	32	8.6%	1	2	32	8.6%
1460	+		+		+		3	0	3	3	50.0%	0	1	5	16.7%	2	3	1	83.3%
1527	+	+	+				3	1	3	10	28.6%	0	1	13	7.1%	0	4	8	42.9%

Outliers: Table 1 shows the top outlier providers for the Sample Provider Group. There are 6 columns (flags) related to 3 outlying scenarios (*Service Levels, Service Intensities* and *Average Amount*), each considered in two versions $OM^m(p)$; where $m = 1$ and $m = 2$, correspond to "local" and "global" outliers respectively. The final score is the total number of flags for each provider.

The second section of Table 1, to the right-hand side, is a colour coded representation of the number of provider/claim combinations identified as outliers based on the given output measure and contained in L^m_{top1} and L^m_{top2}. The individual claims are identifiable from supporting data to this table and consequently targeted for review.

In Fig. 1 below we present the top 7 Sample Provider Group outliers with a total number of 148 provider-claim combinations/points (red dots) and all data points in 2 variables: *Average Amount* (horizontal axis) and *Service Intensities* (vertical axis). Each point in the figure corresponds to one provider-claim combination (20,333 in total, with 6,030 Providers). Each outlier provider is flagged by at least two (out of 4) measures corresponding to *Service Intensities* and *Average Amount* which are considered in two outlying values OM^m, $(m = 1, 2)$. The outlying reason might be "over/under-servicing" and/or "over/under-charging".

This same figure shows that many provider-claims corresponding to the "edge" points (in this 2 dimensional picture) are not in the final outlying list. They are still considered as exceptional cases in their own right however the corresponding provider does not exhibit enough outlying cases (or clients) to be

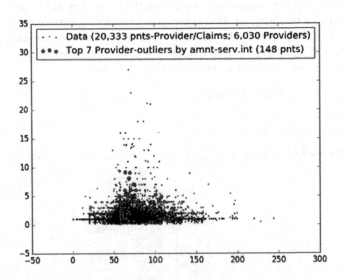

Fig. 1. Top 7 Sample Provider Group outliers with total number of 148 Provider-Claim combinations/points (red dots) and all data points (20,333 in total corresponding to 6,030 Providers) in 2 variables: Average Amount (horizontal axis) and Service Intensities (vertical axis). (Color figure online)

considered a highly ranked outlier overall. Note that in the proposed methodology, the frequency of the provider's claims appearing in the top outlying lists L_{top1}^m and L_{top2}^m is important.

5   Conclusions/Future Applications

This method has proven to be an important and complementary technique to assess the normality of TAC claim treatment by a health/medical provider. When the top 10 outlier providers from Table 1 are compared with SOM output generated from the Sample Provider Group data, only 2 providers appear common between the two techniques. The further 8 outlier providers could be considered complementary and also worthy of review, particularly due to their specific provider-claim combinations/characteristics. Unfortunately, given the subjective nature of assessment and review applied to these outliers, it is neither practicable nor feasible to calculate an accuracy/comparison metric here.

The technique does appear transferable between medical and allied health disciplines with similar treatment patterns (e.g. General Practice, Psychiatry, Psychology, Physiotherapy, Osteopathy, Chiropractic) enabling targeted assessment of treatment levels and subsequent billing amongst peer providers. Additionally, application at a billing provider level could be useful to assess overall outlier behaviour of a clinic or medical centre and not just the treating practitioner. The method will require adaptation to suit discipline specific billing data and additional feature selection will also be available for consideration in alternate disciplines.

Whilst outliers exhibiting "over-servicing" or "over-charging" are interesting to TAC Forensics and the Provider Review Team, provider-claim combinations showing a lack of TAC funded treatment can also be discerned using this method and selected for intervention or indeed linked with specific client outcomes to inform future claims strategy.

Acknowledgements. This paper is the result of project work funded by the Capital Market Cooperative Research Centre in combination with the Transport Accident Commission of Victoria. Acknowledgements and thanks to industry supervisors David Attwood (Lead Research Partnerships) and Marcus Lyngcoln (Manager Forensic Analytics). Additional thanks to Natasha Morphou, Nyree Woods and Gregory O'Neil (TAC Provider Review Team) for their ongoing recommendations and output review.

Appendix

Table 2. Injury types with applicable weights

Order	Type	Class	Weight
1	Fatal	Catastrophic	100.00
2	Quadriplegia	Catastrophic	25.00
3	Severe ABI	Catastrophic	11.00
4	Paraplegia	Other Severe	6.25
5	Brain Injury (Mild)/Head Injury (Ill defined)	Other Severe	4.00
6	Other Spinal	Other Severe	2.78
7	Amputations	Other Severe	2.04
8	Burns (Severe/Moderate)	Other Severe	1.56
9	Internal Injuries	Other Severe	1.23
10	Lost of Sight/Eyes	Other Severe	1.00
11	Degloving	Other Severe	0.83
12	Fractures - Limb	Orthopaedic	0.69
13	Fractures - Other	Orthopaedic	0.59
14	Dislocations	Orthopaedic	0.51
15	Soft Tissue (Neck/Back)/Whiplash	Musculoskeletal	0.44
16	Sprains/Strains	Musculoskeletal	0.39
17	Concussion	Other Injuries	0.35
18	Nerve Damage	Other Injuries	0.31
19	Contusion/Abrasion Laceration	Other Injuries	0.28
20	Other Injuries	Other Injuries	0.25

References

Victorian Transport Accident Commission: Billing Review Program - Information for Medical Practitioners (2016). http://www.tac.vic.gov.au/clients/what-we-can-pay-for/treatment-and-support-services/policy/medical-practitioners/billing-review-program-information-for-medical-practitioners

Australian Government Department of Health: Medicare Benefits Schedule (2016). http://www.mbsonline.gov.au/internet/mbsonline/publishing.nsf/Content/Home

Barnett, V., Lewis, T.: Outliers in Statistical Data, vol. 3, no. 1. Wiley, New York (1994)

Cateni, S., Colla, V., Vannucci, M.: Outlier detection methods for industrial applications. In: Advances in Robotics, Automation and Control. InTech (2008)

Chen, Z., Fu, A., Tang, J.: Detection of outliered patterns. Department of CSE, Chinese University of Hong Kong (2002)

Hawkins, D.M.: Identification of Outliers, vol. 11. Chapman and Hall, London (1980)

Hodge, V., Austin, J.: A survey of outlier detection methodologies. Artif. Intell. Rev. **22**(2), 85–126 (2004)

Ramaswamy, S., Rastogi, R., Shim, K.: Efficient algorithms for mining outliers from large data sets. ACM Sigmod Rec. **29**(2), 427–438 (2000)

Distributed Detection of Zero-Day Network Traffic Flows

Yuantian Miao[1], Lei Pan[2(✉)], Sutharshan Rajasegarar[2], Jun Zhang[1], Christopher Leckie[3], and Yang Xiang[1]

[1] School of Software and Electrical Engineering,
Swinburne University of Technology, Hawthorn, Australia
skymyuanti@gmail.com, {junzhang,yxiang}@swin.edu.au
[2] School of IT, Deakin University, Geelong, Australia
{l.pan,srajas}@deakin.edu.au
[3] Department of Computing and Information Systems,
University of Melbourne, Melbourne, Australia
caleckie@unimelb.edu.au

Abstract. Zero-day (or unknown) traffic brings about challenges for network security and management tasks, in terms of identifying the occurrence of those events in the network in an accurate and timely manner. In this paper, we propose a distributed mechanism to detect such unknown traffic in a timely manner. We compare our distributed scheme with a centralized system, where all the network flow data are used as a whole to perform the detection. We combined supervised and unsupervised learning mechanisms to discover and classify the unknown traffic efficiently, using clustering and Random Forest (RF) based schemes for this purpose. Further, we incorporated the correlation information in the traffic flows to improve the accuracy of detection, by means of using a Bag of Flows (BoFs) based method. Evaluation on real traces reveal that our distributed approach achieves a comparable detection performance to that of a centralized scheme. Further, the distributed scheme that incorporates unknown sample sharing in the framework shows improvement in the zero-day traffic detection performance. Moreover, the classifier used with the combination of BoF and RF shows improved detection accuracy, compared with not using BoFs.

Keywords: Traffic classification · Machine learning
Unknown flow detection · Zero-day traffic

1 Introduction

Network traffic classification plays an important role in network security and management [1–3]. However, the increasing amount of unknown traffic becomes a challenge for effective network traffic classification. In particular, the traffic from zero-day applications (or unknown traffic) forms the major portion of the unrecognized data, accounting for 60% of flows and 30% of bytes, in a network

© Springer Nature Singapore Pte Ltd. 2018
Y. L. Boo et al. (Eds.): AusDM 2017, CCIS 845, pp. 173–191, 2018.
https://doi.org/10.1007/978-981-13-0292-3_11

traffic data set [4]. Not only does the unknown traffic bring difficulties for network management, but it also causes serious security problems. Alazab et al. [5] observed that many obfuscated malware instances that are unable to be detected have eventually resulted in zero-day (unknown) attacks. Hence, it is important to correctly detect the unknown traffic in the network data in a timely manner.

Several machine learning techniques can be employed to address this issue of detecting unknown traffic in the data [6,7]. There are two categories of machine learning algorithms that can be employed to address this challenge, namely supervised and unsupervised algorithms. The supervised algorithms require pre-labelled traffic data (with known traffic classes) to train a classifier [8]. However, the zero-day traffic will not have the labels available for training purposes. The unsupervised algorithms, such as clustering methods, do not require labeled data to find patterns in the traffic flows [9]. Unsupervised learning methods have been employed in the past to detect unknown traffic [5]. However, the accuracy of detection can be improved by combining the supervised and the unsupervised learning schemes to obtain higher purity in the detection of unknown traffic [10]. Further, the amount of traffic that needs to be analyzed is usually high in volume. Hence, using a centralized scheme for detection, where all network traffic data are kept on a centralised sever where the classification process is performed is impractical. Moreover, it delays the detection of new attacks that emerge in the network. Hence, a distributed mechanism is required to address the scalability challenge and to improve the timely detection of zero-day traffic (unknown traffic) in the network.

In this work, we propose a distributed framework with an unknown traffic information sharing mechanism to improve the detection of zero-day traffic in the network. We incorporate both supervised and unsupervised learning mechanisms to improve the accuracy of detection. In particular, we first partition the network traffic into data chunks, and distribute them to several processing nodes. At each node, an unsupervised clustering algorithm is used to cluster the traffic, which has both labeled and unlabeled network traffic flows. Each node then shares the detected unknown traffic data with the other nodes. Each node then performs the unknown-aware classification using the combined network traffic from its own data and the shared information. We compare the performance of three unknown-aware classification methods based on a Random Forest scheme. This process improves the detection accuracy of the unknown traffic in the network in a timely manner.

In our evaluation, we implemented both the centralized and the distributed schemes and performed experiments using a real data set. We used clustering (K-means) as the unsupervised learning algorithm and Random Forest as the supervised learning algorithms for unknown-aware classification. Further, we compared two more variants of the Random Forest based scheme, namely Bag of Flows (BoF) based Random Forest [10], and sub-bag of flow based Random Forest [11] schemes. The results reveal that our distributed scheme achieves comparable detection performance to the centralized scheme with reduced detection delay. Further, we discuss the improvement of overall accuracy of unknown traffic detection achieved for each of the unknown-aware classifiers in detail.

The rest of the paper is organized as follows. Section 2 reviews the related work in detecting and classifying unknown traffic flows. Section 3 explains the centralised and the distributed schemes for unknown traffic detection. Section 4 presents the experiments setup and the results. Section 5 concludes the paper.

2   Literature Review

Network traffic analysis is closely related to various intrusion detection techniques, which is an important field of cyber security applications. In [12], the authors demonstrated that cyber attacks originating from cloud-based machines have become a rapidly increasing issue. There are many clustering based algorithms proposed in the literature to detect anomalous network traffic in the network. In [13], a clustering based algorithm was presented to effectively detect abnormal network traffic in wireless sensor networks.

The real-world challenge arises when the traffic data are encrypted. That is, once the network packets are encrypted, the patterns of the packet contents are disguised so well that it becomes almost impossible to derive reliable signatures based on content analysis. Malicious attackers have been taking advantages of this, hence they use Tor network to evade network traceback [14], advanced encryption schemes like AES to protect the contents of malware [15] and ransomware [16], and so on.

For the above reason, this paper focuses on the analysis of features derived from the statistical patterns of network traffic. That is, we rely on information such as the mean of packet sizes, standard deviation of packet arrival time, median of connection duration time, and so on. In the past few years, there have been many novel systems proposed for unknown traffic detection and classification based on the statistical information of network traffic.

Zhang et al. [10] observed that one of the obstacles towards robust classification is zero-day applications, which contribute unknown traffic in network. They proposed a scheme named Robust statistical Traffic Classification (RTC) to address this challenge. Three key modules within this RTC are unknown discovery, "bag of flows" (BoF)-based traffic classification, and system update. The first unknown discovery module has two steps, which use the K-means algorithm to cluster network flows and then apply Random Forests to further classify the unknown clusters. In the second step, BoF-based traffic classification is considered, that uses the flow correlation information introduced in [17]. In this scheme, the input consists of a few flows rather than the complete traffic flow for constructing the training set. The third step is to update the system with the knowledge of new zero-day traffic. In the classification step, they applied Random Forests, correlation-based classification, semi-supervised clustering and one-class SVM for comparison. It is shown that the RTC performed best among these classifiers for detecting unknown traffic. However, this scheme performs the detection in a centralized fashion. In this paper, we extend this scheme and propose a distributed approach to detect zero-day traffic efficiently with lower computation overhead and in a timely manner.

In 2013, Zhang et al. [18] advocated that a limited amount of information hinders accurate network traffic classification. The authors proposed a new method for the unknown (zero-day) applications problem, which obtained an improved capability on unknown flow detection by utilizing correlation information. Apart from a theoretical analysis of the classification benefit, the comprehensive performance evaluation of the ISP and WIDE (http://mawi.wide.ad.jp/mawi/) data sets showed that the proposed method outperformed existing methods. The proposed system combined flow correlation information into Erman's [19] semi-supervised learning, which can deal with the problem of limited numbers of per-labelled data availability and limited numbers of unknown traffic flows available in the data. In this system, they utilized K-means to construct a Nearest Cluster based Classifier (NCC). A cluster-class mapping is constructed for the testing process of traffic classification. In addition, a flow label propagation is proposed, which is a key process used to increase the size of the labelled traffic data set based on the flow level statistical properties. They also developed a compound classification method combined with "bag-of-flows" by gathering flow predictions from a semi-supervised classifier with the majority vote rule.

Alazab et al. [5] in 2011 addressed the issue of detecting zero-day malware using supervised learning algorithms. They employed robust classifiers and evaluated their performance, including Naive Bayes, k-Nearest Neighbor, sequential minimal optimization with four kernels, backpropagation neural network, and J48 decision tree. Comparing the benign code and malware with the API sequences in a signature database they generalized a training set including the malware traffic. Training the classifier with supervised learning, with 10-fold cross validation, significantly improved the performance. They achieved a true positive rate of more than 98.5% and false positive rate of 2.5% in their evaluation.

Wang et al. [20] focused on automatically deriving reliable signatures from unidentified traffic flows. Their method relied on combining statistic-based flow clustering with payload-based signature matching. This process deals with the problem of insufficient pre-labelled training data set availability. The results show that the signature classifiers constructed from clustering traffic data and pre-labelled traffic data achieved high accuracy. In particular, more than 99% accuracy is achieved. They used feature selection before performing clustering. The method used the Chi-square Ranking algorithm (CHI) with the top five ranked features retained, and two subset-search algorithms, namely Correlation-based Feature Selection (CFS) and Consistency-based Subset Search (CON). For the unsupervised learning to detect unknown traffic, they applied X-means, which has two important parameters, namely the number of clusters and the maximum number of rounds of iteration.

Zhang et al. [17] proposed a Traffic Classification using Correlation (TCC) approach to classify traffic flows and explore their correlation information. There are three new classification approaches introduced in this paper transferred from NN classifiers — AVG-NN, MIN-NN and MVT-NN. With a comprehensive

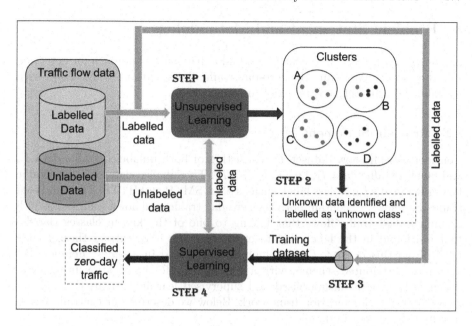

Fig. 1. Zero-day traffic classification framework. Step 1 and Step 2 form the "unknown discovery" phase and the Step 3 and Step 4 form the "unknown-aware classification" phase. Step 1: Traffic flow data that consists of labelled and unlabelled data are processed using a unsupervised learner (K-means clustering algorithm). Step 2: The clusters are labelled based on the majority of the label of the traffic flows it contains. If a cluster does not contain any prelabelled traffic flow data, then that cluster is labeled as the *unknown cluster*. All the traffic flow data in the unknown cluster are then labelled as belonging to a class called *"unknown class"*. Step 3: This unknown class data and the original labelled data are combined to form a labeled training dataset. Step 4: An unsupervised learner (Random forest and its variants) uses the training data to learn the classifier. The learned machine classifies the original unlabelled dataset and identifies the zero-day traffic (i.e., those traffic flows classified as belonging to unknown class). (Color figure online)

comparison, the TCC approach showed better performance than the NN algorithms and has shown the best overall performance compared with C4.5, Erman's clustering-based method and a Bayesian Network. The correlation information utilized the "bag-of-flow" (BoF) traffic classification in their work.

All the above schemes have concentrated on a centralised approach to detect unknown traffic. However, a distributed approach has not been considered to improve the computational complexity and timely detection of zero-day traffic in the network. In this work, we address this limitation by proposing a distributed solution.

3   Problem Statement and Methodology

In this section we present a centralized and a distributed approach for detecting zero-day traffic. We define the zero-day traffic flows as those traffic flows with an unknown (unidentified) protocol (or application).

3.1   Problem Statement

Consider a traffic flow dataset T_c, consisting of both unlabeled traffic data T_u and labeled traffic data T_l, i.e., $T_c = T_l \cup T_u$. The labelled dataset T_l has traffic flows identified with a class label, such as SSH, SMTP and HTTP. The unlabeled traffic data flows T_u contain both the zero-day traffic (unknown samples) as well as unidentified traffic flows that belong to one of the known classes (labels) that are found in the labelled dataset T_l. Our aim is to extract the zero-day traffic from this dataset T_c with high accuracy, and in a timely manner. We propose a distributed scheme, with a zero-day traffic classification framework, which utilises both unsupervised and supervised learning. Figure 1 shows our zero-day traffic classification framework. Below we describe our centralized and distributed schemes in detail.

3.2   Centralised Scheme

The centralized scheme for detecting zero-day traffic is shown in Fig. 2. It contains four parts, namely flow construction, feature extraction, unknown discovery and unknown-aware classification. Flow construction and feature extraction are necessary for statistical-based network traffic classification. Once the network traffic flows are computed, and the features are extracted, an unknown discovery process and unknown-aware classification are performed. For this, we used a framework that we proposed in our previous work [10] to detect the zero day traffic with high purity as shown in Fig. 1. There are two phases involved in this process.

Phase 1: Unknown discovery: Consider the unlabeled traffic data T_u and the labeled traffic data T_l. The unlabeled traffic data T_u contains both the zero-day traffic as well as the unidentified traffic flows that belong to one of those known classes (labels) that are found in the labelled dataset T_l. We aim to extract the zero-day samples from T_u using a semi-supervised approach in the first phase. The labelled (T_l) and unlabelled (T_u) datasets are combined to form a combined dataset $T_c = T_l \cup T_u$. Then clustering (unsupervised) is performed on the combined data T_c to partition them into a set of k clusters $C = \{C_1, C_2, \ldots C_k\}$. We used k-means clustering for this purpose, although any clustering algorithm can be used in practice. We consider that the zero-day traffic are only present in the unlabeled data set T_u. Each cluster in the set C is labeled based on the class label of the majority of the traffic flows that are contained within that cluster. If a cluster does not contain any prelabelled traffic flow data, then that cluster is labeled as the *unknown cluster* or zero-day cluster. All the traffic flows falling within an unknown cluster are labeled as belonging to a new class called the *unknown class*. For example, in Fig. 1, there are four clusters shown, namely, A, B, C and D, where the red and green points represent the unknown and known traffic flows respectively. Clusters A, B and C will be labelled as *known clusters*, and all the traffic flows that fall within a cluster will be given the class label of the majority of the traffic flows it contains. The cluster D, which has only red points in it, will be labeled as the *unknown cluster*. Then, all the traffic flows inside cluster D will be labeled as *unknown class*.

Phase 2: Unknown-aware classification: In this phase, the labeled dataset T_l and all the data from the unknown clusters (Fig. 1) are combined to form a combined dataset T_s. Note that all the traffic flows that fall inside an *unknown cluster* are considered as belonging to a new class called the *"unknown class"*. Hence, the combined data T_s will have traffic flows that are all labled (with known classes and "unknown class"). This labeled T_s is used as the training data to learn a supervised classifier. We use the Random Forest classifier (and other variants as detailed in the Evaluation section) for this purpose, as we have observed in [11] that the Random Forest scheme gives a good classification performance among various other classifiers for network traffic classification. Then the learned Random Forest classifier is used to classify the unlabeled traffic data flows T_u to obtain a high-purity set of zero-day traffic samples.

The above centralised scheme has limitations in terms of scalability with the increase in the amount of network traffic data, and timely detection of zero-day traffic. Hence, a distributed approach is proposed to address these limitations. Below we present our proposed distributed approach in detail.

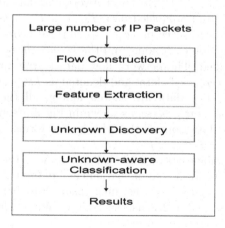

Fig. 2. Centralised scheme for zero-day traffic classification.

3.3 Distributed Scheme

Figure 3 shows the schematics of our distributed approach. This scheme has three nodes, namely Node 1, Node 2 and Node 3, with each one receiving the network traffic flow data identified as `IP Packets 1`, `IP Packets 2`, and `IP Packets 3`, respectively. Apart from the input data sets and the *unknown sample sharing* step, the rest of the four steps are the same as those found in the centralized approach (Fig. 2). The steps involved in the distributed process are as follows.

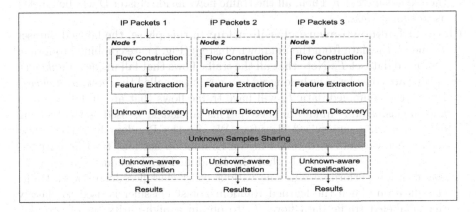

Fig. 3. Distributed scheme for zero-day traffic classification. Node 1, Node 2 and Node 3 are the three distributed processing nodes.

- Each node N_i, $i = 1, 2, 3$ performs the flow construction and feature extraction on its own network traffic data (*IP Packets i, i = 1, 2, 3*) to produce the network flow data T_{ci}, $i = 1, 2, 3$. This network flow data at each node consists of labeled data T_{li} and unlabeled data T_{ui}, i.e., $T_{ci} = T_{li} \cup T_{ui}, i = 1, 2, 3$. Then the *unknown discovery* process is performed using the (unsupervised) clustering algorithm on each node, and the *known clusters* and the *unknown clusters* are identified. The network flow data in the unknown clusters are labeled as *"unknown class"* data, and denoted as $IPi_{uk}, i = 1, 2, 3$.
- All the unknown class data are shared among all the nodes during the *unknown sample sharing process*, as shown in Fig. 3. Each node, on receiving the unknown class data from all the other nodes, forms a combined *unknown class* data $IP_{alluk} = IP1_{uk} \cup IP2_{uk} \cup IP3_{uk}$.
- Each node then combines its own labeled data T_{li} with the combined *unknown class* data IP_{alluk} to form a combined labelled dataset $T_{si}, i = 1, 2, 3$.
- Each node performs supervised learning (using Random Forests), using the combined data $T_{si}, i = 1, 2, 3$ as the training data. This is known as the *unknown-aware classification* process.
- Then the learned Random Forest classifier at each node is used to classify its respective unlabeled traffic flows data T_{ui} as zero-day traffic (unknown class) or not.

Furthermore, it is observed in [17,18,21] that the classification performance and computational complexity can be further improved if the correlation information between the network flows are taken into consideration. Thus, a "bag of flows" (BoF) model is proposed in [17] that factors in this correlation information and applies them in the supervised learning process. The main idea of BoF is as follows. Each BoF contains a few traffic flows that are correlated and come from the same application (class). The traffic flows collected during a certain time period and sharing the same three-tuple (destination_ip, destination_port, transport_protocol) are formed as one BoF. Then, these BoFs are used as the data instead of the individual traffic flows for classification. When one of the BoFs is predicted as belonging to one of the classes, all the traffic flows inside that BoF are predicted as belonging to that same class. In other words, by packaging these traffic flows according to the key correlation information in classification, the prediction error is minimised [17].

In the distributed scheme given above, the unknown-aware classification process uses the Random Forest (RF) algorithm. Based on the advantages observed in incorporating the BoF model for classification in [10], we also experimented with combining the Random Forest algorithm with the BoF algorithm in this work, and call it as *BoF-RF*. In this BoF-RF scheme, the BoF gets the class label of the majority of the class labels predicted for the traffic flows in the BoF by the RF scheme. In addition, we also experimented with another way of combining the RF with a variant of BOF, called sub-bag of flows [10], and we denote it is as *BoF1-RF*. A sub-bag conatains flows sharing 4-tuples: source IP, destination IP, destination port, and transport protocol. Using the sub bags, the BoF gets the class label as follows. A single flow is selected from each sub-bag (as a representative of each sub-bag's flows), and then all the selected flows are classified using the RF. The BoF gets the class label as the label of the majority of those sub-bags' class label. This has an advantage in terms of speeding up traffic classification without significant loss in accuracy.

Among the various steps involved in our centralised zero-day traffic identification and classification framework, unknown discovery is the most time-consuming part, as it has to process the whole input traffic flows at once. As for the distributed system, the amount of input samples used for clustering at each node is significantly smaller than that of the centralized system. Further, the unknown discovery processes can be run simultaneously at various nodes in the distributed system. Hence, the time complexity of the distributed algorithm is relatively smaller than the centralised one.

In practice, the network traffic can change dynamically, and new zero-day traffic can emerge in the network arbitrarily. This means that the processing of the network traffic needs happen dynamically. Our distributed framework can be run periodically, using a pre-defined periodicity (a user defined parameter), in order to capture the newly emerging zero-day traffic in the network. For example, in the context of the ISP data set (that we used in our evaluation), our system can be refreshed weekly, since the network traffic data has been collected during a seven day period. An interesting question here is how this process can be done incrementally? We left this topic for future research.

4   Evaluation

The aim of the evaluation is to compare the performance of zero-day (unknown) traffic detection using the distributed and the centralised scheme. Further, we compare the performance using different classification schemes, namely Random forest, BoF_RF and BoF1_RF schemes. We used a real dataset, called the ISP dataset, for our evaluation [10]. The ISP data set is a trace that we captured using a passive probe at a 100-Mb/s Ethernet edge link from an Internet service provider located in Australia. Full packet payloads are preserved in the collection without any filtering or packet loss. The trace is 7 days long and began on November 27, 2010. After the collection, the ISP data set was preprocessed by a deep packet inspection (DPI) tool, in order to label each flow based on their signature information obtained in the payload contents.

We used F-Measure, precision and recall as the performance metrics for comparison. The corresponding equations for these measures are listed below:

- Overall accuracy is the ratio of the number of correctly classified traffic flows over all testing flows, which measures the classifier from the overall system perspective [18].

$$Accuracy = \frac{\sum(correctly\ classified\ flow)}{\sum(all\ testing\ flows)} \tag{1}$$

- F-Measure is determined using precision and recall, which measures the classifier performance from the per-class perspective [18].

$$F\text{-}Measure = \frac{2 \times precision \times recall}{precision + recall} \tag{2}$$

$$Recall = \frac{TP}{TP + FN} \tag{3}$$

Here, TP represents the true positives and the FN represents the false negatives.

4.1   Experiment Setup

Table 1 shows the 21 statistical features extracted from the ISP data set for use in the evaluation. Note that we use these statistical features for computing the 3-tuple heuristic information for BoF. Thus, knowing a port or the transport protocol information cannot uniquely identify a single traffic flow in the data.

These 21 features are divided into five groups, namely, packet quantity, byte volume, packet size statistic information, inter-packet time statistic features and duration. Since the traffic flow is bidirectional, the flows are typically grouped into two general types: from mobile to server and from server to mobile. Further, the duration between two unidirectional connections is also regarded as one characteristic of each flow.

Table 1. 21 statistical features from ISP data

Type	Feature	Count
Packets	Number of packets transferred from mobile terminal to server	1
Packets	Number of packets transferred from server to mobile terminal	1
Bytes	Volume of bytes transferred from mobile terminal to server	1
Bytes	Volume of bytes transferred from server to mobile terminal	1
Packet size	Min., Max., Ave. and variance of packet size from mobile terminal to server	4
Packet size	Min., Max., Ave. and variance of packet size from server to mobile terminal	4
Inter-packet time	Min., Max., Ave. and variance of Inter packet time from mobile terminal to server	4
Inter-packet time	Min., Max., Ave. and variance of Inter packet time from server to mobile terminal	4
Duration	Duration that the server responding	1

In spite of such abundant feature information, the ISP data set we used contains 300k data records with 18 protocols, namely BT, DNS, EBUDDY, EDONKEY, FTP, HTTP, IMAP, MSN, POP3, RSP, RTSP, SMB, SMTP, SSH, SSL2, SSL3, XMPP, and YAHOOMSG. As for its class distribution, we can generally divide them into three groups: large size, average size and small size. The large size of one class of traffic means its proportion is larger than 10%, which includes SSL2, SSL3, SMTP, HTTP and BT. The average size of one protocol represents that their proportions are roughly between 1% and 10%. Examples include SSH, POP3, MSN, IMAP, EDONKEY, DNS, XMPP, FTP and SMB. The rest of the four classes are considered as the small size group, which includes EBUDDY, RSP, RTSP and YAHOOMSG. Among the small size group, the number of traffic flows corresponding to each protocol is less than 1,000.

Since the number of records with protocols in the small size group are too small to influence the results individually, we combined them and name as the "small" class. The class distribution of the ISP data set with the "small" class is shown in Fig. 4. That is, after the "small" class labelling, there are only 15 protocols in total present among the whole ISP data set. Observing this figure, we define six major classes, namely SSH, SSL2, SSL3, SMTP, HTTP and BT. In order to simulate the problem of zero-day traffic application identification, we manually set a few small size classes and one major class as the "unknown" class. In this paper, we defined the "unknown" class constructed by combining the class SMALL, SMB, XMPP, POP3, IMAP, DNS and SMTP. In the ISP data, the unknown samples (in the unknown class) account for less than 8% of the whole data set.

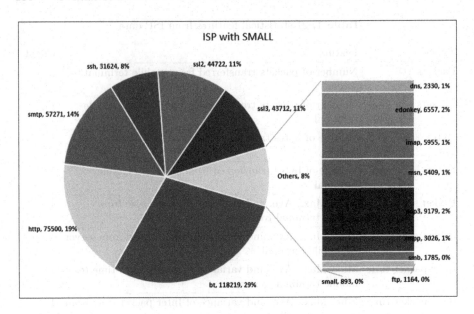

Fig. 4. Class distribution of ISP dataset

In order to arrive at the performance metrics, first, the input data set was divided into three disjointed parts: a pre-labeled set, an unlabeled set, and a testing set. A small percentage of flows were randomly selected from the pre-labeled set and unlabeled set, which were used in the unknown discovery process, and formed the training data set for the final unknown-aware classification. In this evaluation, 1% of the pre-labeled set and 2% of the unlabeled set are used.

As we mentioned in Sect. 3, the unsupervised learning algorithm and the supervised learning algorithm used in the unknown discovery phase are the K-means and the Random Forest algorithms. For the clustering step, the k value of K-means is an important parameter. Zhang et al. [10] studied its influence on zero-day traffic identification and suggested a new optimization method for searching for the optimal k value automatically. This optimization method combines 10-fold cross validation with binary search to find out the optimal k by comparing a key metric, such as the false positive rate (FPR). In this case, the FPR indicates the ratio of the number of flows incorrectly detected as unknown to the whole predicted unknown traffic flows. We performed 10-fold cross validation and repeated the process 10 times to arrive at the classification results.

We report the evaluation of the centralised and the distributed schemes in three views; overall observation, per-class observation and time complexity aspects. For the overall view, overall performance is reported using the overall accuracy. For the per-class observation, we only focus on the zero-day traffic sample, also called the unknown sample for simplicity. The metric used is the F-Measure. For the time complexity measurement, we computed the execution time. It is noted that the three nodes of the distributed system are executed simultaneously. Therefore, the execution time of the distribution system has three values, which differ depending on the various input IP packets used. The whole ISP data set was used as input for the centralized system. The ISP data set was randomly divided into three equal size subsets, which are named as ISP1, ISP2 and ISP3, and used as input for each of the nodes in the distributed scheme.

4.2 Experimental Results

Table 2 presents the *unknown discovery* accuracy along with the best k value selected automatically by the algorithm. The unknown discovery accuracy shown here represents the percentage of traffic correctly discovered as unknown over all the unknown traffic predicted by the unknown discovery phase. In general, all the accuracies observed are over 80%, which demonstrates that the optimization method utilized for selecting k and performing clustering [10] performs well in terms of accuracy. In particular, for the whole ISP data set, the best k value obtained is 512, while the unknown discovery accuracy achieved is 93.90%. The best k chosen for ISP1 is 143 with 97.40% accuracy. In the meantime, the optimized k values and corresponding unknown discovery accuracies obtained for ISP2 and ISP3 are 172 with 89.61% and 141 with 93.16% respectively.

Table 2. The best k value and unknown discovery phase accuracy.

Input data set	Best k value	Unknown discovery accuracy
ISP	512	93.90%
ISP1	143	97.40%
ISP2	172	89.61%
ISP3	141	93.16%

In order to obtain the results from the unknown-aware classification process, we used three algorithms, namely Random Forest (RF), BoF_RF and BoF1_RF. For the distributed scheme, we performed two experiments, one with *sharing* the unknown samples among the other nodes, and the other *without sharing* the unknown samples with the other nodes. Figure 5(a) and the Table 4 show the overall classification performance of the centralized scheme and the first node (node 1) of the distributed scheme using the ISP1 data. In general, the centralized system shows the best overall accuracy among the three experiments. This is expected as the centralised scheme uses the full information from the whole data set for classification. As for the ISP classification, RF performed best against the other two algorithms. As for the distributed scheme with ISP1, apart from the small decrease in accuracy observed for BoF1_RF, the overall accuracy with unknown sharing is slightly higher than the accuracy without sharing the unknown samples discovered by the other two ISP subsets. When compared

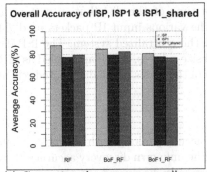

(a) Comparing the average overall accuracy of the whole ISP set after the centralised scheme and ISP1 set after the distributed scheme.

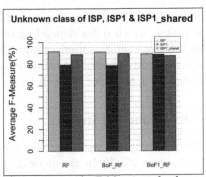

(b) Comparing the F-Measure of unknown class from the whole ISP set after the centralised scheme and ISP1 set after the distributed scheme.

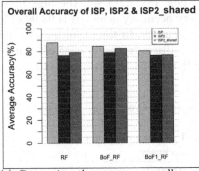

(c) Comparing the average overall accuracy of the whole ISP set after the centralised scheme and ISP2 set after the distributed scheme.

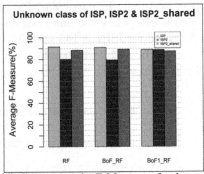

(d) Comparing the F-Measure of unknown class from the whole ISP set after the centralised scheme and ISP2 set after the distributed scheme.

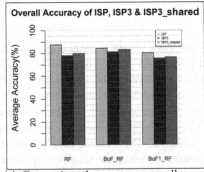

(e) Comparing the average overall accuracy of the whole ISP set after the centralised scheme and ISP3 set after the distributed scheme.

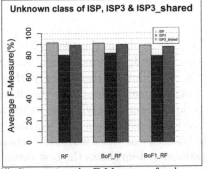

(f) Comparing the F-Measure of unknown class from the whole ISP set after the centralised scheme and ISP3 set after the distributed scheme.

Fig. 5. Comparing the average overall accuracy and F-Measure of unknown class with respect to the three algorithms RF, BoF_RF and BoF1_RF.

with centralized classification and the ISP1 classification with unknown sharing in detail, except for the RF algorithm, the overall accuracy of the latter has declined by two percent compared to the result of the former one. Similar pattern is observed for BoF1_RF classification, where the overall performance from the centralized system is 80% and that from ISP1 with sharing unknown samples is 76%. However, the performance of BoF1_RF from the distributed system is lower than that from the centralized system, which are about 87% and 80% respectively.

As for the per-class observation, the whole ISP with centralized classification also showed the best performance for detecting unknown classes. Figure 5(b) and Table 5 show the accuracy observed for detecting the unknown classes (zero-day traffic) in the data. For the distributed system, classifying ISP1 with unknown sample sharing, the accuracy has improved significantly when classified using the RF and BoF_RF schemes. However, when analysing the performance of the BoF1_RF scheme, it shows better performance in detecting the unknown classes compared with the other algorithms. Its accuracy has slightly declined after sharing the unknown samples. When comparing the results from the centralized classification to that from the distributed classification, the latter's performance is still lower than the former one as expected. However, the differences between the above two performances are small. In particular, using the RF algorithm, the F-Measure of the centralised scheme is around 92%, while that for the distributed scheme with ISP1 is around 88%. Using the BoF_RF scheme, the centralised scheme achieved about 91%, while the distributed scheme showed 89% accuracy. Using the BoF1_RF scheme, the centralized scheme achieved around 89% accuracy, while the distributed scheme achieved 88% accuracy (after sharing the unknown samples).

When we analyse the distributed classification accuracy from node 2, i.e., using ISP2, the results from the overall classification in Fig. 5(c) are quite similar to that from ISP1 classification with unknown sharing. In other words, the overall classification of the centralized system outperforms the distributed system, especially when the classification algorithm used is RF. In contrast to the ISP1 results, the performance of ISP2 with unknown sample sharing, surpassed the results using unknown sharing slightly. In detail, unknown sharing enhanced the RF classifier associated with BoF, that is, the overall accuracy obtained is about 79% before sharing and increased to approximately 83% after sharing. In addition, the unknown sharing technique has also shown positive effect on the RF slightly, with 76% before sharing and 79% after sharing accuracies. As for the BoF1_RF, the performance of ISP2 classification showed slight improvement after sharing the unknown among the nodes.

Figure 5(d) shows the classification results of the ISP and ISP2 for detecting the unknown classes, where the ISP results represent the centralized classification while ISP2 represents Node 2 of the distributed system. The comparison results for F-Measure on the ISP2 data set are quite similar to that of ISP1 data set. It can be noted that after sharing the unknown traffic information, the centralised scheme achieved 91% accuracy, while the distributed scheme (ISP2) with BoF_RF achieved 90%. When observing the effect of unknown sharing in the distributed scheme, the accuracy has improved significantly when used with the RF and BoF_RF schemes. However, this technique has a slightly negative impact (around 1% decline) in the accuracy of ISP2 classification.

Figure 5(e) shows the comparison results among the centralised scheme (ISP), and the distributed scheme at node 3 (ISP3) from the overall perspective. The results are similar to those observed in the previous cases. Overall performance with the BoF_RF scheme are observed to be 85%, 81% and 84% respectively for the ISP, ISP3 without unknown sharing and ISP3 with unknown sharing schemes.

From the per-class view, Fig. 5(f) presents a slightly different result for the unknown class's classification, compared with the previous observations. Though the centralized scheme's (ISP) performance is the best among them, ISP3 with unknown sharing performed well and approximately came closer to the accuracy recorded by the centralized scheme. In particular, when used with the RF classifier, the unknown F-Measure of ISP classification is about 91%, while that of ISP3 with unknown sharing is around 89%. When BoF_RF is used, the unknown F-Measure of ISP is around 90% and that of ISP3 with unknown sharing is approximately to 90%. When used with BoF1_RF, the F-Measure of ISP is roughly 89% while that of ISP3 is 87%. Concerning the unknown sharing technique in the distributed system, it improved the accuracy of the unknown class's classification significantly (around 10% increase).

Table 3 shows the computation time of the centralised scheme and the distributed scheme. It can be observed that the distributed scheme improves the time delay in detecting and classifying the unknown class. The centralised scheme consumed around 4 h to compute the results, whereas each node of the distributed scheme took around 26 min on average to perform the detection.

Table 3. Execution time for each classification.

Experiments	Execution time
Centralized_ISP	4 h
Distributed_ISP1	24 min
Distributed_ISP2	21 min
Distributed_ISP3	27 min

4.3   Discussion

Overall, the evaluation demonstrates that the distributed approach is capable of performing zero-day traffic classification with comparable accuracy to the centralised scheme. Further, the distributed scheme with unknown sample sharing performs better than the one that does not share the information among the nodes. Further, we compared the detection performance using three supervised learning mechanisms, namely Random Forest (RF), Bag of Flows with RF

Table 4. Overall accuracy of the centralised and the distributed schemes

Experiments	RF	BoF_RF	BoF1_RF
ISP	87.61%	84.39%	80.31%
ISP1	77.27%	79.16%	77.25%
ISP1_Shared	79.36%	82.15%	76.28%
ISP2	76.37%	78.80%	76.34%
ISP2_Shared	79.09%	82.41%	76.67%
ISP3	77.98%	81.15%	75.49%
ISP3_Shared	80.01%	83.23%	76.53%

Table 5. F-Measure for unknown class detection in the centralised and distributed schemes

Experiments	RF	BoF_RF	BoF1_RF
ISP	91.40%	90.92%	88.91%
ISP1	78.97%	78.38%	88.48%
ISP1_Shared	88.62%	89.32%	87.30%
ISP2	80.12%	79.32%	88.64%
ISP2_Shared	88.34%	89.13%	87.39%
ISP3	80.10%	81.63%	79.11%
ISP3_Shared	89.07%	89.47%	87.46%

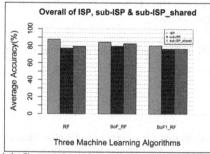

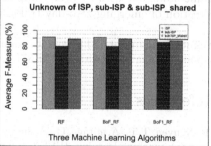

(a) Comparing the average overall accuracy of the whole ISP set using the centralized scheme, distributed scheme with and without shared samples.

(b) Comparing the F-Measure of unknown class from the whole ISP set using the centralized scheme, distributed scheme with and without shared samples.

Fig. 6. Comparison of accuracy and F-Measure.

(BoF_RF) and BoF1_RF schemes. The BoF_RF scheme shows relatively better performance than the RF alone for the distributed scheme. The BoF1_RF scheme showed the lowest performance, although it has an advantage in terms of improved computational complexity. However, the BoF1_RF scheme is slightly faster than the BoF_RF scheme as discussed in Sect. 3.3.

Figure 6(a) and (b) present the overall average performance of the distributed system, computed by combining the output from each node. The results are shown for both the overall view and the unknown class view, where the sub-ISP represents any one node in the distributed scheme.

5   Conclusion

Detecting zero-day traffic is a critical task in order to maintain the security of the network. We present two classification schemes, namely centralized and distributed schemes for detecting such zero-data traffic by combining supervised and unsupervised learning mechanisms. We further analyzed the performance of the distributed scheme with two variants, namely with unknown sample information sharing and without sample information sharing. We used k-means clustering as the unsupervised learning scheme with an automatic mechanism to determine the number of clusters parameter k in our scheme. As the supervised classifier, Random Forest (RF) is used. In order to improve the classification accuracy, we used a Bag of Flows method in addition to the RF scheme, that incorporates correlation information among the traffic flows to perform the classification. Evaluation performed on real traces reveals that our distributed scheme with sharing unknown sample information outperforms the distributed scheme without the sharing information. Further, our distributed algorithm shows comparable accuracy to that of the centralized scheme in terms of detecting the zero-day traffic flows, but with reduced computational overhead, hence facilitating timely detection of emerging zero-day traffic in the network.

Acknowledgement. This work was supported by the National Natural Science Foundation of China under Grant 61401371.

References

1. Nguyen, T.T., Armitage, G.: A survey of techniques for internet traffic classification using machine learning. IEEE Commun. Surv. Tutor. **10**(4), 56–76 (2008)
2. Finamore, A., Mellia, M., Meo, M., Rossi, D.: KISS: stochastic packet inspection classifier for UDP traffic. IEEE/ACM Trans. Netw. (TON) **18**(5), 1505–1515 (2010)
3. Juvonen, A., Sipola, T.: Adaptive framework for network traffic classification using dimensionality reduction and clustering. In: 2012 4th International Congress on Ultra Modern Telecommunications and Control Systems and Workshops (ICUMT), pp. 274–279. IEEE (2012)
4. Kim, H., Claffy, K.C., Fomenkov, M., Barman, D., Faloutsos, M., Lee, K.: Internet traffic classification demystified: myths, caveats, and the best practices. In: Proceedings of the 2008 ACM CoNEXT Conference, p. 11. ACM (2008)

5. Alazab, M., Venkatraman, S., Watters, P., Alazab, M.: Zero-day malware detection based on supervised learning algorithms of API call signatures. In: Proceedings of the Ninth Australasian Data Mining Conference, vol. 121, pp. 171–182. Australian Computer Society, Inc. (2011)
6. Este, A., Gringoli, F., Salgarelli, L.: Support vector machines for TCP traffic classification. Comput. Netw. **53**(14), 2476–2490 (2009)
7. Finamore, A., Mellia, M., Meo, M.: Mining unclassified traffic using automatic clustering techniques. In: Domingo-Pascual, J., Shavitt, Y., Uhlig, S. (eds.) TMA 2011. LNCS, vol. 6613, pp. 150–163. Springer, Heidelberg (2011). https://doi.org/10.1007/978-3-642-20305-3_13
8. Criminisi, A., Shotton, J., Konukoglu, E., et al.: Decision forests: a unified framework for classification, regression, density estimation, manifold learning and semi-supervised learning. Found. Trends Comput. Graph. Vis. **7**(2–3), 81–227 (2012)
9. Hastie, T., Tibshirani, R., Friedman, J.: Unsupervised learning. In: Hastie, T., Tibshirani, R., Friedman, J. (eds.) The Elements of Statistical Learning. Springer Series in Statistics, pp. 485–585. Springer, New York (2009). https://doi.org/10.1007/978-0-387-84858-7_14
10. Zhang, J., Chen, X., Xiang, Y., Zhou, W., Wu, J.: Robust network traffic classification. IEEE/ACM Trans. Netw. (TON) **23**(4), 1257–1270 (2015)
11. Miao, Y., Ruan, Z., Pan, L., Zhang, J., Xiang, Y., Wang, Y.: Comprehensive analysis of network traffic data. In: 2016 IEEE International Conference on Computer and Information Technology (CIT), pp. 423–430. IEEE (2016)
12. Han, Y., Chan, J., Alpcan, T., Leckie, C.: Using virtual machine allocation policies to defend against co-resident attacks in cloud computing. IEEE Trans. Dependable Secure Comput. **14**(1), 95–108 (2017)
13. Rajasegarar, S., Leckie, C., Palaniswami, M.: Hyperspherical cluster based distributed anomaly detection in wireless sensor networks. J. Parallel Distrib. Comput. **74**(1), 1833–1847 (2014)
14. Ling, Z., Luo, J., Wu, K., Yu, W., Fu, X.: Torward: discovery of malicious traffic over Tor. In: 2014 Proceedings IEEE INFOCOM, pp. 1402–1410. IEEE (2014)
15. Conti, M., Mancini, L.V., Spolaor, R., Verde, N.V.: Analyzing android encrypted network traffic to identify user actions. IEEE Trans. Inf. Forensics Secur. **11**(1), 114–125 (2016)
16. Kharraz, A., Robertson, W., Balzarotti, D., Bilge, L., Kirda, E.: Cutting the gordian knot: a look under the hood of ransomware attacks. In: Almgren, M., Gulisano, V., Maggi, F. (eds.) DIMVA 2015. LNCS, vol. 9148, pp. 3–24. Springer, Cham (2015). https://doi.org/10.1007/978-3-319-20550-2_1
17. Zhang, J., Xiang, Y., Wang, Y., Zhou, W., Xiang, Y., Guan, Y.: Network traffic classification using correlation information. IEEE Trans. Parallel Distrib. Syst. **24**(1), 104–117 (2013)
18. Zhang, J., Chen, C., Xiang, Y., Zhou, W., Vasilakos, A.V.: An effective network traffic classification method with unknown flow detection. IEEE Trans. Netw. Serv. Manag. **10**(2), 133–147 (2013)
19. Erman, J., Mahanti, A., Arlitt, M.: QRP05-4: internet traffic identification using machine learning. In: IEEE GLOBECOM 2006, pp. 1–6, November 2006
20. Wang, Y., Xiang, Y., Yu, S.Z.: An automatic application signature construction system for unknown traffic. Concurr. Comput.: Pract. Exp. **22**(13), 1927–1944 (2010)
21. Zhang, J., Chen, X., Xiang, Y., Zhou, W.: Zero-day traffic identification. In: Wang, G., Ray, I., Feng, D., Rajarajan, M. (eds.) CSS 2013. LNCS, vol. 8300, pp. 213–227. Springer, Cham (2013). https://doi.org/10.1007/978-3-319-03584-0_16

False Data Injection Attacks in Healthcare

Mohiuddin Ahmed$^{(\boxtimes)}$ and Abu S. S. M. Barkat Ullah

Canberra Institute of Technology, Reid, ACT 2601, Australia
{mohiuddin.ahmed,abu.barkat}@cit.edu.au
https://cit.edu.au/

Abstract. False data injection attacks (FDIA) are widely studied mainly in the area of smart grid, power systems and wireless sensor networks. In this paper, an overview of the FDIA is proposed including the definition and detection techniques proposed so far. The main focus of this paper is to create awareness about the impact of the FDIA in domains other than smart grid such as healthcare. The impact of FDIA in healthcare is overlooked for last couple of years around the globe. However, the recent information security incidents rise in the healthcare sector reaffirms the requirements of preventive measures for FDIA in healthcare. In this paper, we also focus on the emerging attacks on the healthcare domain to understand the importance of FDIA prevention techniques.

Keywords: FDIA · Smart grid · Healthcare

1 Introduction

Information Technology has a great impact on social wellbeing, economic growth and national security. However, it is also embraced by a group of people with malicious intent, known as cyber criminals. Cyber security has become an integral part in any organization and the mass usage of networked systems has given rise to critical threats such as zero-day vulnerabilities which has a significant financial and social impact [1–4]. Despite research in the area of cyber security increased significantly, are yet to be mitigated [5].

False data injection is known for its severe impact and one of the widely studied cyber attacks in smart grid, power systems, control systems, SCADA networks etc [6]. In layman's terms, it is a type of cyber attack where the compromised sensors reflect manipulated events. Supervisory Control and Data Acquisition (SCADA) systems are widely used for monitoring and control of Industrial Control System (ICS) of national critical infrastructures, including the emerging energy system, transportation system, gas and water systems, and so on. Due to being critical infrastructure, the FDIA in smart grid/SCADA systems has attracted a great deal of attention in recent years.

Unlike smart grid, healthcare has become one of the prime targets for the cyber attacker as it's impact is unprecedented and profitable [7,8]. The impact

© Springer Nature Singapore Pte Ltd. 2018
Y. L. Boo et al. (Eds.): AusDM 2017, CCIS 845, pp. 192–202, 2018.
https://doi.org/10.1007/978-981-13-0292-3_12

of FDIA in healthcare has not been researched before and it is envisaged that it will have a tremendous impact on healthcare arena as the incorporation of sensor networks are on the rise [9–12]. A successfully launched FDIA will lead to incorrect decision making and relevant unwanted actions which may have significant repercussions such as mistreating patients, wrong diagnostics and so on. Therefore, healthcare domain is vulnerable to FDIA and the main contribution of this paper is to:

- Showcase research in FDIA and
- Report current healthcare sector cyber incidents and
- Create awareness for FDIA in healthcare.

In summary, this key objective of this paper is to raise awareness to consider preventive measures for FDIA in healthcare assuming the impact will be saving lives of millions.

1.1   Paper Organization

Rest of the paper is organized as follows. Section 2 contains a simple taxonomy on FDIA research wit hits definition. Section 3 discusses top attacks and reasons for cyber attacks in healthcare. Section 4 contains a critical analysis on FDIA and healthcare followed by conclusion in Sect. 5.

2   FDIA Research

SCADA systems are widely used for monitoring and control of national critical infrastructures, including the emerging energy system, transportation system, gas and water systems, and so on [13]. In general, ICS is comprised of Programmable Logic Controllers (PLCs), Remote Terminal Units (RTUs) with Intelligent Electronic Devices (IEDs), a telemetry system, a Human Machine Interface (HMI) and a supervisory (computer) system. The smart grid can be seen as a large cyber-physical system, made up of the modern communication, information and control technology. Smart grid takes the advantage of the advanced sensor and measuring technology and decision support system to regulate energy and power accurately, ensuring a dependable, secure and efficient power grid. False data injection attack is one of the predominant attacks in the smart grid and can have severe impact on the critical infrastructures. In [14,15] it is shown that, an attacker can exploit the configuration of a power system to launch FDIA to add arbitrary errors into certain state variables while bypassing existing techniques. The pioneer in FDIA research provided a mathematical model for false data injection in smart grid [14,15] as below.

2.1 Conceptual Framework of FDIA

While defining the FDIA [14,15], the following assumptions are made:

- Real measuring vector, $R\{r_1, r_2, \ldots . r_m\}$
- Observed measuring vector, $R_a\{r_{a_1}, r_{a_2}, \ldots . r_{a_m}\}$
- False data vector, $a\{a_1, a_2, \ldots . a_m\}$
- $R_a = R + a$; when $a \neq 0$.

In case of the false data injection attack, the a is not a zero vector. Though, this definition is originally devised for smart grid, it can be seen from the healthcare perspective as long as the healthcare domain is now heavily dependent on sensor networks [16]. In Sect. 4 of this paper, we show that malware is a type of FDIA which is predominantly affecting the healthcare industry.

2.2 A Taxonomy of FDIA Countermeasures

Figure 1 portrays a simple taxonomy of the FDIA countermeasures, especially the prominent detection and prevention techniques. In the scope of this paper, we showcase a wide variety of FDIA countermeasures in the smart grid domain. The following subsections contains a brief description of different types of FDIA countermeasures. However, all of these approaches are applicable in the smart grid applications and can be customized for healthcare domain.

GLRT for Colored Gaussian Noise. [17] considered the problem of false data injection detection in smart grid when the measurements are corrupted by colored Gaussian noise. They developed a model for the noise with the autoregressive process. Then, estimated the state of the power transmission networks and developed a generalized likelihood ratio test (GLRT) detector for identifying the false data injection attacks. The proposed GLRT system was evaluated on the IEEE 30-bus power system with comparison to conventional Gaussian noise based detector.

Spatio-Temporal Correlations. According to [18], most current research on FDIA focuses on countermeasures for traditional power grids rather smart grid infrastructures. Unlike traditional approaches, they developed an efficient and real-time scheme to detect FDIA in smart grids, by exploiting spatial-temporal correlations between grid components. This solution focuses on the spatiotemporal cyber-state correlations and trust-based voting to evaluate the reliabilities of state estimations. The attacks are detected once those unreliable state estimations are identified. They conducted the experimental analysis on US smart grid consisting of 48 states and found reliable solution.

Hop-by-Hop Authentication. Smart grid encompasses the sensor networks which are often deployed in unattended environments, and therefore vulnerable to false data injection attacks. According to [19], the standard authentication approaches cannot prevent this FDIA when the attacker has compromised one

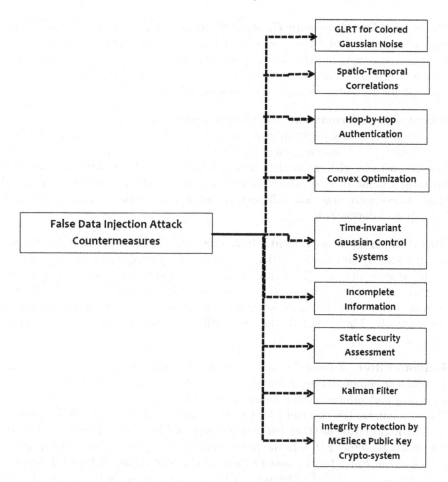

Fig. 1. A taxonomy of FDIA countermeasures

or a small number of sensor nodes. To detect FDIA in sensor networks, they developed three interleaved hop-by-hop authentication schemes that guarantee that the base station of the sensor network is able to detect injected false data immediately if a certain number of nodes are compromised. In addition, the proposed schemes enable an en-route node to detect and drop injected false data instantly to save energy.

Convex Optimization. The potential financial risks as a consequence of the FDIA are explored by [20]. The key issues addressed here, are the problems of the data injection attack against state estimation in deregulated electricity markets, as an initiator of financial misconduct. The authors provided a heuristic for finding profitable attack, formulated as a convex optimization problem. The potential attacks are demonstrated on an IEEE 14-bus system.

Time-Invariant Gaussian Control Systems. A false data injection attack model is proposed and the effects on a linear time-invariant Gaussian control system is analyzed in [21]. It is proven that the existence of a necessary and sufficient condition under which the attack could destabilize the system while successfully bypassing a large set of possible failure detectors.

Incomplete Information. A realistic false data injection attack is essentially an attack with incomplete information due to the attackers lack of real-time knowledge [22]. A mathematical model is provided to characterize false data injection attacks with incomplete information from both the attacker's and grid operator's point of view. Additionally, a vulnerability measure is introduced that can compare and rank different power grid topologies against FDIA with incomplete information.

Static Security Assessment. It is shown by [6] that the SSA results are manipulated by launching a FDIA. Two types of targeted scenarios are proposed in the paper which are fake secure signal attack and fake insecure signal attack. The former attack will deceive the system operator and the latter will deceive the system operator to take unnecessary actions, such as generator rescheduling, load shedding, etc. The proposed analysis is validated with the IEEE-39 benchmark system.

Kalman Filter. Kalman Filter (*'a set of mathematical equations that provides an efficient computational (recursive) means to estimate the state of a process, in a way that minimizes the mean of the squared error'*) [24] was adopted by [23] to minimize the effect of FDIA in smart grid. The Kalman Filter estimates and the information are then fed into the proposed Euclidean detectors, that has the ability to detect FDIA in the power system. The χ^2-detector is incorporated with Kalman Filter for the measurement of the relationship between dependent variables and a series of predictor variables. The analysis in this paper showed that the χ^2-detectors are unable to detect FDIA while the Euclidean distance metrics can identify the same.

Integrity Protection by McEliece Public Key Crypto-System. A FDIA prevention technique presented by [25] based on protecting the integrity of the measurements at measurement units and during their transmission to the Control centres. The proposed scheme alleviates the negative impacts of FDIA on grid's performance using the McEliece public key crypto-system. Experimental analysis shows that the prevention scheme guarantees the integrity and availability of the measurements with lightweight overhead.

3 Cyber Attacks in Healthcare

In this section, we showcase the top healthcare cyber attacks and the issues responsible for these attacks. The main objective of this section is to create the relationship between the FDIA in smart grid and healthcare. We also discuss the impact of cyber security in Australian healthcare industry.

3.1 Most Visible Cyber Attacks in Healthcare

In this section, we briefly describe the most significant attacks in healthcare.

Stolen Financial Data: Anthem. In 2015, cyber criminals took control of personal information for 80 million patients and stole tens of millions of records [26]. It is considered as one of largest data breaches of healthcare information discovered in history. The breach or attack is believed to be executed by a well-resourced cyber espionage group, known as Black Vine. The attackers used custom-developed malwares. These malwares are known as 'Hurix', 'Sakurel' and 'Mivast'.

Insurance Fraud: Community Health Systems. According to US Department of Justice (DoJ), Community Health Systems [28] reported that hackers had stolen patient data from 4.5 million individuals [27]. A sophisticated malware software that took advantage of a test server lacking the proper security features in place for Internet connectivity, was used to copy patient data of any patient that had received services or been referred for services. Because the appropriate security measures were not in place, hackers were able to locate VPN credentials found within the test server and then log into CHSs infrastructure.

Ransomware: Presbyterian Medical Center. The hackers used malware to infect the computers at Hollywood Presbyterian Medical Center in February 2016 [29–31]. They demanded a ransom of $17,000 to restore their operation. After investigation it is found that an employee opened an infected email or downloaded the malware from a pop-up ad which facilitated the virus to be planted in the network. According to 2017 Verizon Data Breach report [32], ransomware surged from the 22nd most common type of malware in 2014 to the fifth most common this year.

Currently, there are two different types of ransomware. The first is through a program named 'Locky'. 'Locky' exploits spam email campaigns where an email is sent across the system that contains infected MS Word documents. Once the victim opens the document, macros will install on the host computer and begin infecting the network. The second ransomware is via the 'Samas' program, which attacks web servers directly.

Social Engineering: University of Washington Medicine. Social engineering has become a common trend to deploy malware to infect healthcare systems. Hackers target organizations that publicly display their employee's contact information [33]. Hackers sent phishing emails [34] containing links or attachments that appear to be innocent in nature. However, once the link is accessed, it will immediately infect the user's computers and begin to spread throughout the rest of the health system.

Using phishing techniques, hackers compromised nearly 90,000 patients at University of Washington Medicine had their personal information in 2013.

An employee was sent an email that had a malicious link embedded into the content as an attachment and when opened, malware took over the computer and accessed the employee's computer, which contained files needed for billing patients.

MEDJACK: UCLA Health. A new type of hijacking in healthcare is known as medical device hijack (called MEDJACK) [35]. The attacker places malware within the networks through a variety of methods such as malware laden website, targeted email, infected USB stick, socially engineered access, etc. This allows backdoors which gives access to cyber-criminals for months before detected. Data can be easily stolen as nothing abnormal appears in the network.

A recent victim of MEDJACK is UCLA Health, where personal data for 4.5 million patients were exposed. The personal data included names, birth dates, Medicare numbers, and health plan numbers.

3.2 Top Reasons for Cyber Security Risks in Healthcare

In this section, we highlight the top reasons for increased security risks in healthcare.

High Demand for Medical Records in the Black Market. The high demand for patients medical records in the black market is alluring cyber-attackers [37]. Electronic health records (EHRs) include names of patients, their birth dates, policy numbers, diagnosis codes, and billing information. Fraudulent users exploit these data in many ways, such as creating fake IDs to buy medical equipment or medications that can be resold. According to the Federal Bureau of Investigation, (EHR) are more valuable than financial data. EHRs can sell for $50 in the black market, compared to just $1 for a stolen social security number or credit card number.

Bring Your Own Device (BYOD) Policy. Healthcare companies are encouraging physicians, nurses, and other medical staff to bring their personal devices such as tablets, smart phones, and laptops to work [38]. It is considered as another reason for exploiting vulnerabilities in the organizations network. Additionally, in the era of IoT (Internet of Things), it is creating more chances for the cyber criminals to exploit any network.

Limited Spending. Spending to ensure healthcare information security has been underwhelming [36]. However, it is estimated that investments in the industry against cyber attacks will only reach $10 billion worldwide by 2020. Unfortunately, it is under 10% of the total expenditure on critical infrastructure security. Due to insufficient budget, healthcare security countermeasures are not as effective as other domain such as smart grid.

Employee Negligence. Although cyber-attacks can be considered as the leading cause of data breaches in the healthcare industry, there are still many other security issues that were caused by negligent employees. An employee, for example, may open an email attachment that contains malware and compromise confidential information stored in a computer.

3.3 Healthcare Cyber Attacks and Predictions in Australia

Australia's most significant healthcare cyber attack, exposed more than one million personal and medical records of Australian citizens donating blood to the Red Cross Blood Service [39]. In Australia, the healthcare sector accounted for 27% of attacks in 2016. Compared to financial, which was only 12%, it is reflected that Australians medical records are of significant value to the hackers.

The growing risk of cyber attacks leaves the Australian economy exposed to a potential $16 billion damage according to one of the world's biggest insurance companies [40]. In a joint study with Cambridge University, the Lloyd's insurance company found that, out of 301 global cities, Sydney ranks 12th in terms cyber-attack exposure with $4.86 billion of economic growth at risk. According to City Risk Index 2015–2025, Lloyd's says Sydney is the riskiest Australian city, followed by Melbourne, Brisbane, Perth, Adelaide and Canberra.

4   FDIA in Healthcare: Prevention Is Better Than Cure

The research on FDIA is focussed on smart grid domain. However, the components of smart grid such as sensor networks, measuring equipments are also used in healthcare domain. Additionally, we have discussed the usage of malware for healthcare cyber attack in previous section. Malware is considered as a way of FDIA. Therefore, the FDIA can also be exploited in healthcare sector. It is important to raise concern over the impact of FDIA in healthcare, because human lives are at stake in this case unlike smart grid. FDIA in smart grid may have financial impact, however, the FDIA in healthcare may cause several impacts. We focus on few scenario as below where FDIA can hit and will lead to dangerous consequences.

- **Wrong Diagnosis:** If the medical devices which are part of a sensor network can be compromised, then the injected false data can be leading to wrong diagnosis. For example, if the medical device which measures the blood pressure of a patient is fed with false data, then the diagnosis based on the manipulated blood pressure will lead to wrong/unwanted treatment. Therefore, the patient may face deteriorated health conditions.
- **Illegal Insurance Claim:** Injecting false medical records may cause the insurance provider to pay unnecessarily as the patient data is not legitimate. For example, if cyber attacker, injects data for a surgery which is suppose to be covered by insurance, then the beneficiary will get paid without under going any surgery!

- **Mission Critical Factors:** While a complicated surgery is going on and the physicians are dependent on the real time health data (blood pressure, pulse, temperature etc.), even a slight variation of these data (as a consequence of FDIA) will lead to loss of life.

Based on the points above, we can understand the repercussion of FDIA in healthcare. Therefore, it is imperative and requires immediate attention to develop countermeasures for FDIA In healthcare. As it is always better to prevent than cure.

5   Conclusion

This paper is the intersection of FDIA in smart grid and healthcare. FDIA is focussed on smart grid operations however, it is also applicable for healthcare industry due to the similarity in infrastructures. We showcased the FDIA research in smart grid and presented the recent cyber incidents in healthcare. Then tried to establish a relationship between FDIA and healthcare domain. In summary, in this paper, we tried to create awareness by showcasing different FDIA countermeasures in smart grid and increasing cyber incidents in the healthcare industry. We are optimistic that, researchers in the cyber security domain will have more attention in developing FDIA countermeasures for healthcare as these may save millions of lives rather than only financial loss.

References

1. Ahmed, M., Mahmood, A.N., Hu, J.: A survey of network anomaly detection techniques. J. Netw. Comput. Appl. **60**, 19–31 (2016)
2. Ahmed, M., Mahmood, A.N., Hu, J.: Outlier detection, Chap. 1. In: The State of the Art in Intrusion Prevention and Detection, pp. 3–21. CRC Press, New York, January 2014
3. Ahmed, M., Mahmood, A.N., Islam, M.R.: A survey of anomaly detection techniques in financial domain. Future Gener. Comput. Syst. **55**, 278–288 (2016)
4. Ahmed, M., Mahmood, A.N.: A novel approach for outlier detection and clustering improvement. In: 2013 8th IEEE Conference on Industrial Electronics and Applications (ICIEA), pp. 577–582, June 2013
5. Buczak, A.L., Guven, E.: A survey of data mining and machine learning methods for cyber security intrusion detection. IEEE Commun. Surv. Tutor. **18**(2), 1153–1176 (2016)
6. Chen, J., Liang, G., Cai, Z., Hu, C., Xu, Y., Luo, F., Zhao, J.: Impact analysis of false data injection attacks on power system static security assessment. J. Mod. Power Syst. Clean Energy **4**(3), 496–505 (2016)
7. Healthcare top target for cyberattacks in 2017: Experian predicts. http://www.healthcareitnews.com/news/. Accessed 12 May 2017
8. Cyber Attacks: In the Healthcare Sector. https://www.cisecurity.org/. Accessed 12 May 2017
9. Pachauri, G., Sharma, S.: Anomaly detection in medical wireless sensor networks using machine learning algorithms. Procedia Comput. Sci. **70**, 325–333 (2015)

10. Salem, O., Guerassimov, A., Mehaoua, A., Marcus, A., Furht, B.: Anomaly detection scheme for medical wireless sensor networks. In: Furht, B., Agarwal, A. (eds.) Handbook of Medical and Healthcare Technologies, pp. 207–222. Springer, New York (2013). https://doi.org/10.1007/978-1-4614-8495-0_8
11. Salem, O., Guerassimov, A., Mehaoua, A., Marcus, A., Furht, B.: Sensor fault and patient anomaly detection and classification in medical wireless sensor networks. In: 2013 IEEE International Conference on Communications (ICC), pp. 4373–4378, June 2013
12. Haque, S.A., Rahman, M., Aziz, S.M.: Sensor anomaly detection in wireless sensor networks for healthcare. Sensors 15, 8764–8786 (2015)
13. Ahmed, M., Anwar, A., Mahmood, A.N., Shah, Z., Maher, M.J.: An investigation of performance analysis of anomaly detection techniques for big data in SCADA systems. EAI Endorsed Trans. Ind. Netw. Intell. Syst. 15(3), 5 (2015)
14. Liu, Y., Ning, P., Reiter, M.K.: False data injection attacks against state estimation in electric power grids. In: Proceedings of the 16th ACM Conference on Computer and Communications Security, CCS 1909, pp. 21–32. ACM, New York (2009)
15. Liu, Y., Ning, P., Reiter, M.K.: False data injection attacks against state estimation in electric power grids. ACM Trans. Inf. Syst. Secur. 14(1), 13:1–13:33 (2011)
16. Alemdar, H., Ersoy, C.: Wireless sensor networks for healthcare: a survey. Comput. Netw. 54(15), 2688–2710 (2010)
17. Tang, B., Yan, J., Kay, S., He, H.: Detection of false data injection attacks in smart grid under colored Gaussian noise. In: 2016 IEEE Conference on Communications and Network Security (CNS), pp. 172–179, October 2016
18. Chen, P.Y., Yang, S., McCann, J.A., Lin, J., Yang, X.: Detection of false data injection attacks in smart-grid systems. IEEE Commun. Mag. 53(2), 206–213 (2015)
19. Zhu, S., Setia, S., Jajodia, S., Ning, P.: Interleaved hop-by-hop authentication against false data injection attacks in sensor networks. ACM Trans. Sen. Netw. 3(3), 14 (2007)
20. Xie, L., Mo, Y., Sinopoli, B.: False data injection attacks in electricity markets. In: 2010 First IEEE International Conference on Smart Grid Communications, pp. 226–231, October 2010
21. Mo, Y., Sinopoli, B.: False data injection attacks in control systems. In: First Workshop on Secure Control Systems, CPS Week, pp. 226–231 (2010)
22. Rahman, M.A., Mohsenian-Rad, H.: False data injection attacks with incomplete information against smart power grids. In: 2012 IEEE Global Communications Conference (GLOBECOM), pp. 3153–3158, December 2012
23. Manandhar, K., Cao, X., Hu, F., Liu, Y.: Combating false data injection attacks in smart grid using kalman filter. In ICNC, pp. 16–20. IEEE (2014)
24. Welch, G., Bishop, G.: An introduction to the kalman filter. Technical report, Chapel Hill, NC, USA (1995)
25. Abdallah, A., Shen, X.S.: Efficient prevention technique for false data injection attack in smart grid. In: 2016 IEEE International Conference on Communications (ICC), pp. 1–6, May 2016
26. Healthcare Breach Report 2016. http://www.ciphertex.com/. Accessed 12 May 2017
27. Insurance Fraud: Community Health Systems. https://www.justice.gov/. Accessed 12 May 2017
28. Community Health Systems, Inc. http://www.chs.net/. Accessed 12 May 2017
29. Ransomware: Hollywood Presbyterian Medical Center. http://resources.infosecinstitute.com/category/healthcare-information-security/. Accessed 12 May 2017

30. Why Health Care is Especially Vulnerable to Ransomware Attacks. http://fortune.com/2017/05/15/ransomware-attack-healthcare/. Accessed 12 May 2017
31. Ransomeware   Rising.   http://www.healthcareitnews.com/news/ransomware. Accessed 12 May 2017
32. Why Health Care is Especially Vulnerable to Ransomware Attacks. http://www.verizonenterprise.com. Accessed 12 May 2017
33. 90,000 Patients Compromised at UW Medicine. https://www.infosecurity-magazine.com/. Accessed 12 May 2017
34. Ahmed, M., Kaysar, J.: Phishing attack protection-PAP-approaches for fairness in web usage. Int. J. Comput. Sci. Issues 8(6), 258–261 (2011)
35. MEDJACK: UCLA Health. http://pulse.embs.org/may-2016/healthcare-hacked/. Accessed 12 May 2017
36. Healthcare organizations boost spending on cyber security. https://betanews.com/2017/02/21/healthcare-security-spending/. Accessed 12 May 2017
37. High demand for medical records in the black market. https://www.promisec.com/industries/health-care/. Accessed 12 May 2017
38. Morrow, B.: BYOD security challenges: control and protect your most sensitive data. Netw. Secur. 2012(12), 5–8 (2012)
39. Cyber attacks: a growing risk for healthcare professionals. https://securitybrief.com.au. Accessed 12 May 2017
40. The World's specialist insurance market. https://www.lloyds.com/. Accessed 12 May 2017

Identifying Precursors to Frequency Fluctuation Events in Electrical Power Generation Data

Md. Shahidul Islam$^{(\boxtimes)}$, Russel Pears, and Boris Bačić

School of Engineering, Computer and Mathematical Sciences,
Auckland University of Technology, Auckland 1010, New Zealand
`sshahid01921@gmail.com`,
`{russel.pears,boris.bacic}@aut.ac.nz`

Abstract. To predict an occurrence extraordinary phenomena, such as failures and fluctuations in an electrical power system, it is important to identify precursor events that signal an impending fluctuation event. In this paper we integrate wavelet analysis with statistical inference methods to identify a precursor pattern for frequency fluctuation prediction. The frequency time series data was converted into the wavelet domain to extract the time dynamics after which change point detection methods were used to signal significant deviations in the wavelet domain. The change points extracted were taken as early indicators of a fluctuation event. Using historical data on known fluctuation events we trained a regression model to estimate the gap between a change point and its corresponding fluctuation point. Our results show that change points could be predicted a number of time steps in advance with a low false alarm rate.

Keywords: Frequency fluctuation · Change point detection · Precursors
Wavelet transform · Correlation and regression

1 Introduction

In time series prediction, precursor pattern identification is a key mechanism for indicating future events of interests in a system. In an electrical power generation system a slight imbalance between electricity demand and generation can cause system frequency to fluctuate. Typically, electrical grid operators are required to ensure that the system frequency is stable around a nominal frequency generation value (for example, 50 Hz in New Zealand and many other countries). Maintenance of frequency within the normal band and management of time error is managed through frequency regulating reserves. Frequency regulating reserve is the provision of spare synchronized generating unit capacity with a response time fast enough to control frequency within the normal band of 49.85 to 50.15 Hz.

The stability and reliability of the electricity grid is dependent on efficient design and management of the power system [1]. Frequency management is critical to operation, as many system components have operating ranges within a narrow frequency band. Any sufficiently large frequency deviation will isolate important system components,

© Springer Nature Singapore Pte Ltd. 2018
Y. L. Boo et al. (Eds.): AusDM 2017, CCIS 845, pp. 203–219, 2018.
https://doi.org/10.1007/978-981-13-0292-3_13

generators, and consequently disconnect load. This represents a world-wide problem but is more acute in countries such as New Zealand due to its reliance on nuclear-free and renewable sources of energy [2].

The electricity sector in New Zealand uses mainly renewable energy sources such as hydropower, geothermal power and increasingly, wind energy. Electricity is produced in both North and South islands and interchange takes place via the High Voltage Direct Current (HVDC) link across the two islands [3]. In the context of frequency fluctuations in electrical grids there are two types of events which are of interest: firstly, when the signal value tracks above the upper tolerance limit. In this case, the operator needs to put in place measures to decrease generation levels in order to bring the signal back to its normal band. The second type of event is the opposite, whereby the signal tracks below the lower tolerance limit. In this situation, the load exceeds the generated power levels and the operator needs to use power reserves to increase the level of generated power [4]. Thus, from a computational perspective the objective is to predict the timing and class of a fluctuation event, given the recent history of the frequency signal. We shall show in our empirical study that analysis of the most recent time series segment that spans activity from the last known fluctuation point is sufficient to predict the next fluctuation event with high accuracy.

There are a number of studies reported on precursor detection in many different application areas including detection of abnormal ECG activity, precursors to earthquakes, patterns leading to traffic accidents, etc. Some of these studies exploited domain specific knowledge. In this study we focus on wavelet transform [5] and change point detection methods [6] to identify precursor patterns that signal an impending fluctuation event. Our results show that localized patterns, captured in the time domain serve as accurate predictors of fluctuation events. In this respect, wavelet analysis played a fundamental role in capturing local dynamics of a time series sequence.

2   Research Objectives

The main objectives of the research are as follows:

i. To estimate the occurrence of fluctuation events with high true positive and low false positive rates. The maximization of true positive rates will ensure that fluctuation events are missed with low probability, thus enabling counter measures to be put in place. Low false positive rates will ensure that the system stability is not affected by regulating power generation levels in anticipation of a false fluctuation.

ii. To perform multiple step ahead to enable system operators to put in place measures to mitigate or prevent frequency fluctuations given knowledge of the likely timing of such fluctuation events.

3   Literature Review

There have been a number of previous studies in precursor pattern identification from time series data. Among them signal processing methods such as Fourier analysis and Wavelet analysis have been widely used. Wavelet transforms have been used for detection of geomagnetic precursors of earthquakes [7]. Alperovich et al. [8] and Gwal et al. [9] used wavelet analysis to identify precursors for seismic and seismogenic ULF emissions.

Some researchers attempted to detect precursory events of earthquakes based on pre-defined precursory patterns [10]. Others used an autoregressive time series approach based on AR modelling [11]. Hallerberg et al. [12] explain their process to identify precursors by using AR (1) and evaluate the success of predictions via receiver operator characteristics curves.

More recently, the problem of precursor detections are discussed in terms of symbolic dynamics and it has been applied for detecting earthquake precursors [13] and failure precursor in electrical systems [14]. In order to detect and monitor failure precursors and anomalies early in electrical systems, they have developed a signal processing method that can detect and map patterns to an anomaly measure. Batch-style scenarios, such as Key Graph [15] and based clustering [16], have been proposed for finding risky active faults.

In the financial domain, sawtooth precursor activity [17] in ASDEX was analysed by using a band power correlation method combined with a wavelet transform.

Also in medical science arena, the wavelet transform was used to identify precursors. Saritha et al. [18] deals with the study of ECG signals using wavelet transform analysis. In the first step an attempt was made to generate ECG waveforms by developing a suitable MATLAB simulator and in the second step, using wavelet transform, the ECG signal was denoised by removing the corresponding wavelet coefficients at higher scales. Robert et al. [19] also used wavelet analysis for determining precursors of health abnormalities from processing medical records.

There has also been extensive research in the related area of change point detection. Pears et al. [20] present a novel approach to the concept change detection problem. They propose two change detectors, SeqDrift1 and SeqDrift2 that have significantly better false positive rates and also use the Bernstein bound [21] for detecting changes within the change detection window. Moskvina and Zhigljavsky [22] used the singular spectrum analysis (SSA) technique for change detection in time series, based on the SVD of the Hankel matrix. Terumasa et al. [23] used extension of the singular spectrum analysis (SSA) technique to identify precursor patterns.

4   Research Methodology

Our methodology is based on transforming the raw time series data collected at the 30 s resolution level into wavelet form in order to correlate fluctuations with patterns in the wavelet spectrum. As described in the literature review the wavelet transform has been used successfully to identify precursors in other application domains such as detection of ECG abnormalities and hence our focus on the wavelet transform. Out of the many different types of wavelet transforms available the Morlet was selected as it was designed to work with continuous data.

The Morlet wavelet is used to model the dynamics of the time series and change detection is used to detect volatility in the Morlet wavelet domain. The dataset spanning six months is first divided into segments, with the start of a segment as the observation that occurred just after the previous sequence's known fluctuation point and the end of the sequence as the last observation in which the sequence is in a fluctuation state. We then process each sequence with the Morlet wavelet.

At first we apply the Morlet wavelet transform on our raw time series data and generate wavelet coefficients. Secondly, by using sliding window approach we move step by step forward and by taking a threshold value we identify high wavelet coefficients in every step of sliding window. We then calculate density proportion of the high coefficients and also calculate rate of change of density proportion between two successive windows. The Hoeffding bound is then used for change detection over the sequence. We compare the rate of change of the sequence with the Hoeffding bound and create a change point. Finally, we train a regression model from ground truth data on fluctuation points using historical data. The regression model is used to predict the timing of the fluctuation beyond the occurrence of the change point.

Our overall methodology can be expressed by the following diagram (Fig. 1):

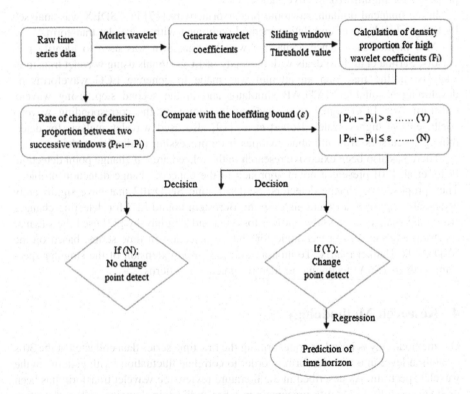

Fig. 1. Overall research methodology

We now briefly cover the main tools used in our methodology.

4.1   Wavelet Transform

Wavelet Transform is a mathematical function used to divide a given function or continuous-time signal into different scale components. Usually one can assign a frequency range to each scale component. Each scale component can then be studied with a resolution that matches its scale. A wavelet transform is the representation of a function by wavelets. The wavelets are scaled and translated copies (known as "daughter wavelets") of a finite-length or fast-decaying oscillating waveform (known as the "mother wavelet").

Wavelet analysis is localized both in time domain and frequency domain, different from square wave analysis of time domain and Fourier analysis of traditional frequency domain. Wavelet analysis equals to a windowing technique with variable-sized regions; it allows the use of long time intervals where we want more precise low frequency information, and shorter regions where we want high frequency information. Therefore, wavelet analysis can focus on and analyze any detail of studied signal [24]. The function form of wavelet transform is given by

$$\Phi_{a,b}(t) = \frac{1}{\sqrt{a}} \Phi\left(\frac{t-b}{a}\right) \tag{1}$$

Where, $\Phi_{a,b}$ is the Wavelet function, a is scale and b is position (time) in wavelet transform.

Let x(t) be the input signal and Φ be the chosen wavelet function, the continuous wavelet coefficient of x(t) at scale a and position b is:

$$C_{a,b} = \int x(t) \frac{1}{\sqrt{a}} \Phi\left(\frac{t-b}{a}\right) dt \tag{2}$$

Where, $C_{a,b}$ is the Continuous wavelet function with scale a and position (time) b.

The Morlet wavelet or Gabor wavelet [25] is a wavelet composed of a complex exponential (carrier) multiplied by a Gaussian window (envelope). This wavelet is closely related to human perception, both hearing and vision. Morlet wavelet equation is defined by

$$\Phi(t) = \pi^{-1/4} \cos(kt) e^{-t^2/2} \tag{3}$$

Where, k is the wave number and $k = e^{-1/2}\sigma^2$.

More generally the expression of Morlet wavelet as follows:

$$\Phi(t) = \left(1 + e^{-\sigma^2} - 2e^{-\frac{3\sigma^2}{4}}\right)^{-\frac{1}{2}} \pi^{-\frac{1}{4}} e^{-\frac{t^2}{2}} \left(e^{i\sigma t} - e^{-\frac{\sigma^2}{2}}\right) \tag{4}$$

Where, $\Phi(t)$ is the Morlet wavelet at point t.

4.2 Hoeffding Inequality

Our approach to change detection relies on well-established bounds for the difference between the true population and sample mean of a given data sample. A number of such bounds exist that do not assume a particular data distribution. Among them are the Hoeffding, Chebyshev, Chernoff and Bernstein inequalities. In probability theory, Hoeffding's lemma is an inequality that bounds the moment-generating function of any bounded random variable. Hoeffding's inequality provides an upper bound on the probability that the sum of independent random variables deviates from its expected value [26]. In contrast, the Hoeffding inequality provides a better bound and is thus adopted in our work. The Hoeffding inequality states as follows:

Let $X1, X2, \ldots, Xi$ are independent random variables taking values in the interval $[ai, bi]$ with mean μ. Then for all > 0,

$$\Pr\left(\left|\frac{1}{n}\sum\nolimits_{i=1}^{n} Xi - \mu\right| > \varepsilon\right) \leq 2\exp\left(\frac{-2n^2\varepsilon^2}{\sum_{i=1}^{n}(b_i - a_i)^2}\right) \tag{5}$$

In case of unit interval $[0, 1]$ with mean μ, the Hoeffding inequality states the following:

$$\Pr\left(\left|\frac{1}{n}\sum\nolimits_{i=1}^{n} Xi - \mu\right| > \varepsilon\right) \leq 2\exp(-2n\varepsilon^2) \tag{6}$$

4.3 Correlation

Correlation is any of a broad class of statistical relationships involving dependence, though in common usage it most often refers to the extent to which two or more variables have a linear relationship with each other [27]. Correlation coefficients are used in statistics to measure how strong a relationship is between two or more variables. The quantity r, called the linear correlation coefficient, measures the strength and the direction of a linear relationship between two variables. Correlations are useful because they can indicate a predictive relationship that can be exploited in practice [28]. The linear correlation coefficient is sometimes referred to as the Pearson product moment correlation coefficient in honor of its developer Karl Pearson. It's denoted by small r and the mathematical formula for computing r is:

$$r = \frac{n\sum xy - (\sum x)(\sum y)}{\sqrt{n(\sum x^2) - (\sum x)^2}\sqrt{n(\sum y^2) - (\sum y)^2}} \tag{7}$$

Where n is the number of pairs of data samples.

4.4 Regression

In statistical modeling, regression analysis is a statistical process for estimating the relationships among variables. It includes many techniques for modeling and analyzing several variables, when the focus is on the relationship between a dependent variable and one or more independent variables [29]. More specifically, regression analysis helps one understand how the typical value of the dependent variable (or 'criterion variable') changes when any one of the independent variables is varied, while the other independent variables are held fixed [30]. Most commonly, regression analysis estimates the conditional expectation of the dependent variable given the independent variables – that is, the average value of the dependent variable when the independent variables are fixed. Regression analysis is widely used for prediction and forecasting. It is also used to understand which among the independent variables are related to the dependent variable, and to explore the forms of these relationships. A linear regression equation is usually written

$$Y_i = \beta_0 + \beta_1 X + \epsilon \tag{8}$$

Where, Y = the dependent variable, X = the independent variable (or covariate), β_0 = the intercept term, β_1 = the slope or regression coefficient of the model, ϵ = the error term.

5   Results and Discussion

The frequency time series for a six-month period spanning 16 October 2014 to 01 April 2015 is shown in Fig. 2. Data was collected at a resolution of 30 s. Figure 2 shows fluctuations in the frequency signal takes place continuously in time but fluctuations outside a given tolerance band are considered to be potentially harmful to grid infrastructure and to consumer equipment. In New Zealand, the grid operator specified the tolerance band to be in range [49.85, 50.15] Hz. Fluctuations outside this band are referred to as error events and these are highlighted in Fig. 2. As Fig. 2 shows that error events are a common occurrence and thus a mechanism to identify them in advance is important to the reliable operation of the electrical grid.

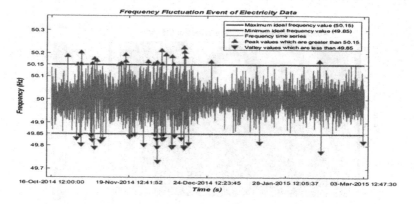

Fig. 2. Frequency time series with error events

Our methodology is to analyze known error events present in historical data and attempt to identify features that will predict future occurrences of such error events. In order to achieve this objective, we divided the dataset into distinct segments with each sequence spanning a time period between two successive error events. In order to make statistically reliable predictions we ensure that each sequence contains at least 60 time points, shorter sequences were discarded. A total of 981 sequences were obtained and subjected to analysis. Figure 3 shows a sample of 3 sequences obtained in this manner.

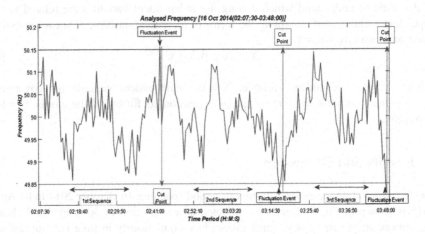

Fig. 3. Frequency time series into sequences

The Morlet wavelet algorithm was then applied to each segment separately and a 2D wavelet spectrum in the form of a scalogram spanning 32 levels of resolution. Figure 4 shows a distinctive signature that was repeated across all 981 sequences we analysed. The red region represents high coefficient values while the blue region consists of low coefficients. The red regions on the extreme left coincides with time

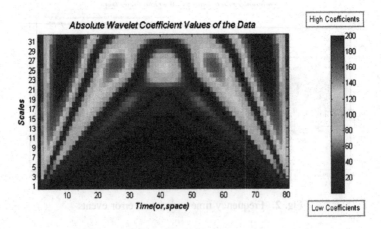

Fig. 4. 2D figure of wavelet transform on raw data (Color figure online)

points that occur just after the previous error event while the red region on the extreme right coincides with time points that occur just before the next error event. The blue region in the middle is associated with time points where the frequency signal is relatively stable with fluctuations occurring well within the tolerance band. Given that this basic pattern repeats across all sequence analysed this strongly suggests that high coefficient regions are precursors to error events.

We start by taking the first 40 data points and then applying the wavelet transform. Thereafter, by taking interval 5 like 1–45, 1–50 and so on, we generate similarly wavelet pattern in each stage of a sequence. Figure 5(a), (b), (c), (d), (e), (f), (g) shows the result and it indicates that when we move forward step by step then high coefficient region (Red region) as well as number of wavelet coefficient is changing on the right region for every length of a sequence with time points.

Now by using a sliding window approach we move step by step forward in intervals of size 5 on the wavelet coefficient array, producing windows such as 1–40, 6–45, 11–50 and so on. Using a threshold value on wavelet coefficients, we identify high wavelet coefficients in every instance of a sliding window. We then calculate the density proportion of the high coefficients and also calculate the rate of change of density proportion between two successive sliding windows. We compare the rate of change of the sequence with the Hoeffding bound ε which is calculated from Hoeffding inequality and create a change point. If rate of change of density proportion would be more than the ε then we call a change point detect on that situation. More precisely, if we get

$$|P_{i+1} - P_i| \leq \varepsilon; \quad \text{No change point}$$
$$|P_{i+1} - P_i| > \varepsilon; \quad \text{Change point}$$

Where, P_i = Density proportion of high wavelet coefficient of 1st window.

$$P_i = N_H/N_T$$

N_H = Number of high wavelet coefficients in the right region.
N_T = Total number of coefficients in the right region.
P_{i+1} = Density proportion of high wavelet coefficient of next window.
ε = Hoeffding bound calculated from Hoeffding inequality.

The detection of a change point implies that an abnormality exists at a certain point on the sequence and from that point we assume that a fluctuation will occur shortly afterwards. Our result shows that at a certain point the rate of change between two successive density proportions becomes very high and it exceeds the threshold value obtained by applying the Hoeffding bound. By applying our methodology on historical data, we then in a position to pair off each change point with the actual fluctuation point, thus producing a training dataset for building a regressor for prediction on time series sequences that arrive in the future. Get a change point before the fluctuation event and stored it our library. In our experimentation we noted that change points always occur in the red region of the wavelet coefficient spectrum.

In building our regression model we consider the rate of change of density proportion of high wavelet coefficient region as an independent variable (X) and the length of size of the data between change point and just before the fluctuation event (Y), here in after referred to as the gap, as the dependent variable. Figure 6 shows the procedure for defining our variables.

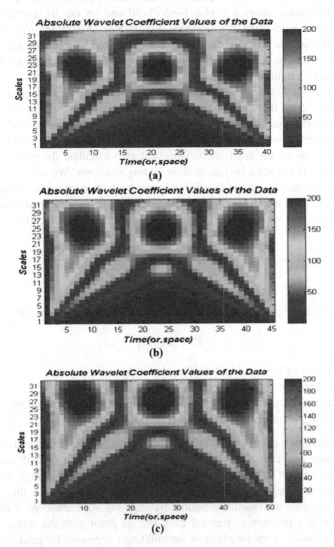

Fig. 5. (a): Wavelet transform on first 40 data points. (b): Wavelet transform on 1–45 data points. (c): Wavelet transform on 1–50 data points. (d): Wavelet transform on 1–55 data points. (e): Wavelet transform on 1–60 data points. (f): Wavelet transform on 1–65 data points. (g): Wavelet transform on 1–70 data points. (Color figure online)

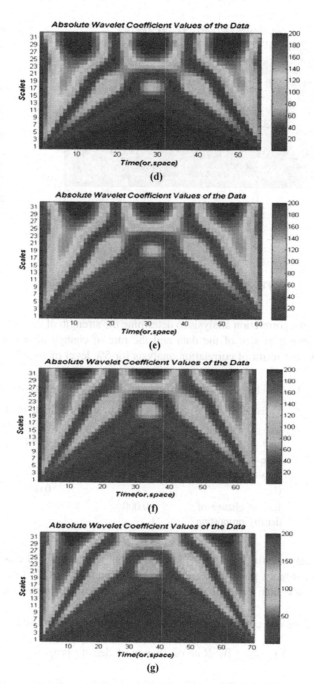

Fig. 5. (*continued*)

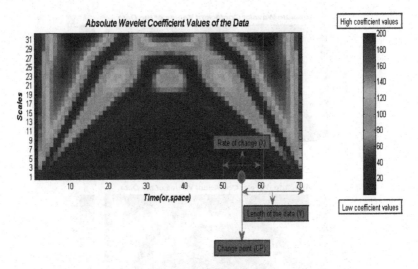

Fig. 6. Feature extraction and variable definition

We first run a correlation analysis to determine the strength of the linear relationship between the length of size of the data and the rate of change of density proportion. Table 1 shows the result of correlation analysis for 5% level of significance.

Table 1. Correlation between gap and rate of change of density proportion

Correlation	Variables	Length of size of the data	Rate of change of density proportion
Pearson correlation (r)	Gap	1	0.864
	Rate of change of density proportion	0.864	1
Significance (two tailed)	Gap	-	0.000
	Rate of change of density proportion	0.000	-

The correlation result (r = 0.864) suggested that gap and rate of change of density proportion are highly positively correlated. This implies that, if rate of change of density increases, then we will obtain a bigger gap for the prediction of frequency fluctuation.

Now we consider a regression model for the prediction purpose as follows:

$$\text{Gap(Y)} = \beta_0 + \beta_1 \text{ [Rate of change of density proportion(X)]} \qquad (9)$$

Table 2 shows the summary result of regression analysis. It shows that the R-square (R^2) value is 0.747, which representing that 74.7% variation of the dependent variable (gap) is due to variation in the independent variable (rate of change of density proportion), which in fact, is a strong explanatory power of regression.

Table 2. Regression model summary

Model	R square	Adjusted R square	Std. error of the estimate	F test	P value
(9)	0.747	0.746	0.770	2890.2	0.000

Predictor: rate of change of density proportion.

Dependent variable: gap between change point and just before the fluctuation event. In Table 3, unstandardized coefficients indicate how much the dependent variable varies with an independent variable, when the independent variable is held constant. The beta coefficients indicate how and to what extent the rate of change of density proportion influences the gap between the change point and the fluctuation event.

Table 3. Coefficients of the regression model

Model (9)	Unstandardized coefficients		Standardized coefficients	t value	P value
	Beta (β)	Std. Error	Beta (β)		
Intercept	19.78	0.143	12.194	138.1	0.000
Rate of change of density proportion	−34.45	0.641	−1.323	53.76	0.000

From Table 3, the fitted regression model in case of unstandardized coefficients is given by:

$$\text{Gap(Y)} = 19.78 - 34.45 \text{ [Rate of change of density proportion (X)]} \qquad (10)$$

For the justification of goodness of fit of the regression model we also estimate residuals for regression and draw a normal probability plot of residuals. Figure 7 shows the probability plot of residual for our regression model and from the plot we clearly visualize that our suggested model is well fitted.

Our dependent variable, gap (Y) approximates the normal distribution and its descriptive statistics are shown in Table 4.

To estimate the false positive rate for prediction of fluctuation events we use a statistical estimation procedure. In this respect we define a false positive if the change point (precursor) if the ratio S(i) between the offset of the change point position (CP) from the start of the sequence to the total length (TL) of the sequence is less than 3 standard deviations from the sample mean taken across all sequences. The reasoning here is that the change point is too premature and does not signal the fluctuation event with sufficient time proximity.

$$S(i) = \frac{CP}{TL}; \ i = 1, 2, 3, \ldots n \text{ (Sequence number)}$$

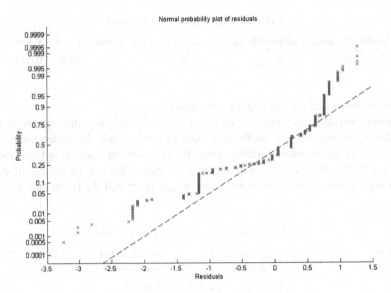

Fig. 7. Normal probability plot for residuals, u_i

Table 4. Descriptive statistics for the gap (Y) distribution.

Total number of sequences	Maximum value	Minimum value	Range	Mean	Standard deviation
981	15	9	6	12	1.5306

In order to compensate for the different sized sequences we normalized the S(i) variable. We compute the sample mean (μ) and standard deviation (Ơ) of S. We then use the following condition:

$$S\,(i) \geq \mu - 3(\text{Ơ})$$

If any value of S(i) is higher than the [μ − 3(Ơ)] then we signal a false positive.

On the other hand if the fluctuation occurs before a change point is detected then a false negative event has occurred. As we had ground truth we are able to record the false positive and true positive (1-false negative) rates. In our analysis of 981 sequences we obtained a false positive rate less than 3% at the 5% level of significance. At the same time we were able to achieve a perfect true positive rate as we were able to detect a change point prior to the fluctuation event for every single sequence, thus giving a true positive rate of 100% at the 5% level of significance.

6   Conclusion

Our research shows that the Morlet wavelet is an effective mechanism for extracting precursors to frequency fluctuations in electrical time series data. The Morlet wavelet was used to generate coefficients and power spectrum from our raw data and in every sequence we obtained a high coefficient red region which has a high dominant power within the entire sequence. After calculating rate of change of density proportion of wavelet coefficient we consistently obtained a change point in the position of the red region of the wavelet coefficient spectrum. From the change point signalled by this region, we are in a position to make a prediction that a frequency fluctuation will occur in the near future. The regression model that we built with rate of change as the predictor proved to be highly successful in predicting the gap between the change point detected and the fluctuation point. In effect, the change point was the precursor to the fluctuation that occurs downstream in the data.

On average (across the 981 sequences we analysed in the 6-month period), the fluctuation was predicted to occur 12 data points away from the change point, this giving the grid operator 6 min to put in place measures to deal with the impending fluctuation. In future work we plan to widen the gap by mapping a newly arriving sequence to its nearest neighbour which is stored in a library containing past sequences whose fluctuations are known. We plan to employ the Dynamic Time Warping (DTW) method to implement the mapping as the two time series to be matched could potentially be of significantly different lengths and have widely different time dynamics.

Acknowledgments. We would like to thank Mike Phethean and Nabil Adam, Transpower NZ Ltd for supplying the data and explaining to us the complexities of power management on the New Zealand national electrical grid.

References

1. Anon: Electricity authority-the act, code and regulations, Electricity Authority of New Zealand. Technical report (2010). www.ea.govt.nz/act-code-regs/
2. New Zealand's Energy Outlook, Electricity Insight: Exploring the uncertainty in future electricity demand and supply. A Publication of the Ministry of Business, Innovation and Employment (2014). ISSN 1179-3996 (print), ISSN 1179-4011 (online)
3. Mortensen, L.: New Zealand Transmission Grid (2012)
4. Alexandra, V.M.: Electric Power Systems: A Conceptual Introduction. IEEE Press/Wiley Interscience, New York (2006)
5. Peng, Y.H.: Wavelet and Its Application to Engineering. Science Press, Beijing (1999). (in Chinese)
6. Vidakovic, B.: Statistical Modeling by Wavelets, Probability and Statistics. Wiley, New York (2000)
7. Alperovich, L., Zheludev, V.: Wavelet transform as a tool for detection of geomagnetic precursors of earthquakes. J. Phys. Chem. Earth 23(9–10), 965–967 (1998)

8. Alperovich, L., Zheludev, V., Hayakawa, M.: Use of wavelet analysis for detection of seismogenic ULF emissions. Radio Sci. **38**(6), 1093 (2003). https://doi.org/10.1029/2002RS002687

9. Gwal, A.K., Shaheen, R., Panda, G., Jain, K.S.: Study of seismic precursors by wavelet analysis. Res. J. Eng. Sci. **1**(4), 48–52 (2012)

10. Huang, H.M., Fan, H.S., Bian, Y.J., Zou, L.Y.: Investigation into the automatic recognition of time series precursor of earthquakes. Acta Seismol. Sin. **11**(5), 605–614 (1998)

11. Bello, G.D., Lapenna, V., Macchiato, M., Satriano, C., Serio, C., Tramutoli, V.: Parametric time series analysis of geoelectrical signals: an application to earthquake forecasting in Southern Italy. Ann. Geophys. **39**(1), 11–21 (1996)

12. Hallerberg, S., Altmann, E.G., Holstein, D., Kantz, H.: Precursors of extreme increments. Phys. Rev. E **75**, 016706 (2007)

13. Karamanos, K., Peratzakis, A., Kapiris, P., Nikolopoulos, S., Kopanas, J., Eftaxias, K.: Extracting preseismic electromagnetic signatures in terms of symbolic dynamics. Nonlinear Process. Geophys. **12**, 835–848 (2005)

14. Patankar, R.P., Rajagopalan, V., Ray, A.: Failure precursor detection in complex electrical systems using symbolic dynamics. Int. J. Signal Imaging Syst. Eng. **1**(1), 68–77 (2008)

15. Ohsawa, Y.: Key graph as risk explorer in earthquake-sequence. J. Contingencies Crisis Manag. **10**, 119–128 (2002)

16. Fukuda, K., Koganeyama, M., Shouno, H., Nagao, T., Kazuki, J.: Detecting seismic electric signals by LVQ based clustering. In: International Conference on Parallel and Distributed Processing Techniques and Applications, pp. 1305–1311 (2001)

17. Papp, G., Pokol, G., Por, G., Igochine, V.: Analysis of sawtooth precursor activity in ASDEX Upgrade using bandpower correlation method. In: Igochine and ASDEX Upgrade team, 36th EPS Conference on Plasma Phys, Sofia, 29 June–3 July 2009. ECA, vol. 33E, P-1.157 (2009)

18. Saritha, C., Sukanya, V., Narasimha, M.Y.: ECG signal analysis using wavelet transforms. Bulg. J. Phys. **35**, 68–77 (2008)

19. Robert, M.P., Thomas, E.P., Barbara, G.B.: Method and system for determining precursors of health abnormalities from processing medical records, US 20110218823 A1, 8 Sept, UT-Batteiies LLC, Oak Ridge, TN, USA (2011). https://www.google.com/patents/US20110218823

20. Pears, R., Sakthithasan, S., Koh, S.Y.: Detecting concept change in dynamic data streams - a sequential approach based on reservoir sampling. Mach. Learn. **97**(3), 259–293 (2014). https://doi.org/10.1007/s10994-013-54339

21. Sakthithasan, S., Pears, R., Koh, Y.S.: One pass concept change detection for data streams. In: Pei, J., Tseng, V.S., Cao, L., Motoda, H., Xu, G. (eds.) PAKDD 2013, Part II. LNCS (LNAI), vol. 7819, pp. 461–472. Springer, Heidelberg (2013). https://doi.org/10.1007/978-3-642-37456-2_39

22. Moskvina, V., Zhigljavsky, A.: An algorithm based on singular spectrum analysis for change point detection. Commun. Stat.—Simul. Comput. **32**, 319–352 (2003)

23. Terumasa, T., Daisuke, I., Kazuyuki, H., Akimasa, Y., Teiji, U., Akiko, F., Akira, M., Kiyofumi, Y.: Detecting precursory events in time series data by an extension of singular spectrum transformation. In: Selected Topics in Applied Computer Science, pp. 366–374 (2008)

24. Wu, W.: Extracting signal frequency information in time/frequency domain by means of continuous wavelet transform. In: International Conference on Control, Automation and Systems (2007)

25. John, A.: Morlet wavelets in quantum mechanics. Quanta **1**(1), 58–70 (2012). https://doi.org/10.12743/quanta.v1i1.5

26. Hoeffding, W.: Probability inequalities for sums of bounded random variables. J. Am. Stat. Assoc. **58**(301), 13–30 (1963). https://doi.org/10.1080/01621459.1963.10500830. JSTOR 2282952. MR 0144363
27. Cohen, J., Cohen, P., West, S.G., Aiken, L.S.: Applied Multiple Regression/Correlation Analysis for the Behavioral Sciences, 2nd edn. Lawrence Erlbaum Associates, Hillsdale (2003)
28. Rodgers, J.L., Nicewander, W.A.: Thirteen ways to look at the correlation coefficient. Am. Stat. **42**(1), 59–66 (1988). https://doi.org/10.1080/00031305.1988.10475524. JSTOR 2685263
29. Rencher, A.C., Christensen, W.F.: Introduction, Methods of Multivariate Analysis. Wiley Series in Probability and Statistics, vol. 709, chap. 10, sect. 10.1, 3 edn, pp. 19–46. Wiley, Hoboken (2012). ISBN 9781118391679
30. Cox, D.R.: The regression analysis of binary sequences (with discussion). J. R. Stat. Soc. Ser. B (Methodol.) **20**(2), 215–242 (1958). JSTOR 2983890

Social Media and Applications

Social Media and Applications

Collaborative Filtering in an Offline Setting
Case Study: Indonesia Retail Business

Hamid Dimyati$^{(\boxtimes)}$ ⓘ and Ramdisa Agasi ⓘ

Stream Intelligence, Jakarta Selatan 12710, Indonesia
{hamid.dimyati, ramdisa.agasi}@id.streamintelligence.com

Abstract. In the past decade, most modeling efforts to date have been focused on the application of recommender systems in an online setting. However, only a few studies have exclusively addressed the actual challenges that arise from implementing it in an offline system. Although the principles of recommender systems implementation between the online and offline commerce are almost identical to some extent, applying the algorithm in the offline environment has its own unique challenges such as lack of product rating and description. Furthermore, most of the customers in the offline retail tend to purchase favorite products repeatedly in short periods. Overcoming such shortcomings could help offline retail to identify the right product that has a higher likelihood to be purchased by a specific customer, and hence increasing revenue. This paper proposes the use of Item-based Collaborative Filtering algorithm as recommender systems to address the limitations of the offline setting.

Keywords: Recommender system · Collaborative Filtering · Offline commerce

1 Introduction

The demand of the recommender systems has risen to prominence because of the growing demand for product recommendation to a specific customer characteristic or needs. Many online stores had applied recommender systems for a personalized product recommendation. They deployed the recommender algorithm to their site or even created competition to obtain the best algorithm to make the performance of their current recommender systems optimized. The main reason behind such an effort is that the online stores hope to avoid false positive error in which unwanted products that do not match with the interest of customers since the error could lead unsatisfied customers [9].

However, the application of recommender systems should not be limited to the ecommerce environment, as it could be extended to the offline commerce as well. One example of such an application to the offline setting is to increase the effectiveness of promotion and eventually increase revenue. There are three challenges in implementing recommender systems algorithm in the offline retail. First, there is no customer-product rating data. Secondly, there is insufficient product description. Finally, sometimes customers do not want product recommendations they never bought, but they need the same one. Our preliminary study revealed that there is 1.89 times higher chance of customers would buy the same products they ever bought rather than bought different products within six months period. Meanwhile, most of the key motivation behind the

Y. L. Boo et al. (Eds.): AusDM 2017, CCIS 845, pp. 223–232, 2018.
https://doi.org/10.1007/978-981-13-0292-3_14

recommender systems application is that a customer will more likely purchase similar products he/she already bought which is not recommending the exact same products.

The availability of both product description and rating data have already been helpful in building the recommender systems on ecommerce sites. However, in reality, we often face a challenge where the brick and mortar stores lack product details and substantial evidence in capturing the customer preference for a product due to the absence of rating system that provides much clearer representation of customer preference. Those problems will definitely lead to the challenges in applying recommender systems in their environment. Because of that, the most suitable approach is Collaborative Filtering since it does not need any products description. Collaborative Filtering is one of the recommender algorithms that rely on the information about customers' preference to be used in prediction process of what customer will like based on their similarity to others. Meanwhile, lack of ratings (explicit feedback) data in their database will be our next obstruction in implementing Collaborative Filtering to their environment since it does require customers' preference data. Alternatively, we will require any implicit feedback to be used in building the recommender systems. Purchase or buy products can be assumed as implicit feedback [6], and repetition of purchase can be implicit feedback as well [7]. This theory will be a solution in the implementation of Collaborative Filtering within the offline commerce in which their customers have specific behavior to have repeat purchases.

This paper deals with the three main obstacles in building recommender systems for offline setting through the use of Collaborative Filtering for implicit feedback. We applied this recommender system to one of the biggest retailers in Indonesia and obtained success in moving up the marketing effectiveness by achieving the more customers response to their promotion.

The rest of this paper is organized as follows. In Sect. 2, we present any research progress of recommender systems algorithms. Section 3 describes steps that are used to build recommender systems in the offline setting. Section 4 provides the description of data sets and the discussion of our experimental results. Lastly, we present our conclusion of the project and what future potential development of it in Sect. 5.

2 Related Work

There are many techniques to develop recommender systems. In general, there are two algorithms which can be run, they are Collaborative Filtering and Content-based Filtering. The main difference between those two approaches is the information that is collected to measure the similarity between customers or products. Collaborative Filtering uses customers' explicit feedback (ratings) or implicit feedback (purchase history) as their preference of certain products to obtain the product or customer similarities which then be used to deliver the product recommendations that more likely to be purchased by given customers [3]. Meanwhile, Content-based Filtering collects content or description contained in a product or customer as main information to be processed in recommending customers what products that they like based on the product or customer similarities [5].

Each of methodologies has own drawbacks, Collaborative Filtering mainly suffers from cold start problem which is hard to measure the similarity of new products or customers. Another problem is about data sparsity which means that most customers only purchased a very low number of products so that it is hard to measure the similarities between products or customers. On the other hand, Content-based Filtering truly relies on the detail information about products such as its color, size, brand and so forth or customers' demography such as gender, age, hobbies and so on. That information is not always being found in many real-world applications. Considering each shortcoming, Collaborative Filtering seems straightforward to be implemented rather than Content-based Filtering.

Collaborative Filtering has two basic approaches to process customers' preference. Those are a Memory-based and Model-based approach. Memory-based sometimes called Neighborhood-based, requires similarity computation that reflects distance correlation between two customers or products, i and j. There are various similarity measurements that can be used, for example, Correlation-based similarity, Vector Cosine-based, Conditional Probability-based, Jaccard index and many more. On the other hand, the Model-based comprises some kind of models to be used as recommender systems such as Bayesian Belief Net CF Algorithms, Clustering CF Algorithms, Regression-based CF Algorithm, MDP-based CF Algorithms, Latent Semantic CF Models, and other Model-Based CF Techniques [11].

We could expect two alternative output types from Collaborative Filtering which are either predicted feedback or top-N product recommendations. A predicted feedback is to generate predicted rating of products which a given customer has not given any feedbacks. Meanwhile, the top-N recommendation output is to offer a set of N top-ranked products that will be of interest to an active customer. Those two outputs are generated based on products or customers similarities which achieved from customer-product preference. Regarding the output of Collaborative Filtering for implicit preference, it is less applicable to generate predicted feedback because of binary value within the data. It is more relevant to produce top-N recommendation instead. Furthermore, it is crucial to understand any unique natures of implicit feedback such as points below [3]:

1. It is still questionable why customers did not buy a certain product, whether they really dislike the product or because they did not know the product.
2. Implicit feedback is inherently noisy. While we passively track the customer's behavior, we can only guess their preferences and true motives.
3. The numerical value of explicit feedback indicates a preference, whereas the numerical value of implicit feedback indicates confidence.
4. Evaluation of implicit feedback recommender system requires appropriate measures.

3 The Proposed Approach

3.1 Data Preparation

We have a set of customers $U = \{u_1, u_2, \ldots, u_m\}$ and a set of products $I = \{i_1, i_2, \ldots, i_n\}$. The transaction data as implicit feedback is converted into a

$m \times n$ customer-product feedback matrix $R = (r_{jk})$ where every row represents a customer u_j with $1 \le j \le m$ and every column represents product i_k with $1 \le k \le n$. And r_{jk} represent the implicit feedback of customer u_j for product i_k. $r_{jk} \in \{0, 1\}$ where we define

$$r_{jk} = \begin{cases} 1 & \text{customer } u_j \text{ purchased product } i_k \\ 0 & \text{otherwise} \end{cases} \tag{1}$$

Typically, only a small fraction of feedbacks is known and for many cells in R will be missing. This is called as sparsity problem that we will later discuss. This type of data in the context of collaborative filtering has similar situations to classifiers *one-class data* since only the 1-class gives definite meaning which is positive feedback. Differently, the 0-class is a mixture of positive and negative feedbacks [8]. There are two strategies to deal with one-class data that is to assume all missing feedbacks (zero value) are negative feedbacks or to assume that all missing feedbacks are unknown. Here, we assumed that the zero value may reflect negative feedback or customers dislike the products [1].

Thus, we will have customer-product feedback matrix like Table 1.

Table 1. Customer-product feedback matrix example.

	i_1	i_2	i_3	...	i_n
u_1	1	0	1	...	0
u_2	1	1	0	...	1
u_3	1	0	1	...	1
...	...	...	...	...	...
u_m	0	1	0	...	1

Sparsity and Synonymy. As mentioned previously, sparsity problem may happen in the offline setting recommender systems. Although the number of products in the offline commerce is not as many as ecommerce has, the possibility of sparse data is quite high. It comes from too many zero values in the customer-product matrix which reflects that customers tend to buy a few subsets of products [10].

Sparse data could be caused by synonymy within the data. Sometimes, some different products are actually the same products. This may happen due to input error or unclear product hierarchy within the retail. As a result, the recommender systems cannot detect this latent association and obviously will implicate to the performance that may be poor [9].

We propose dimensionality reduction approach to reduce the sparsity and synonymy problems. The approach is by reducing the number of products that have a strong relationship indicated with high Jaccard value. The Jaccard index is a statistic used for comparing the similarity and diversity of sample sets. It is also known as intersection over union. Given two sets of customers defined as: $X = \{u_x\}$ and $Y = \{u_y\}$ where $x, y = \{1, 2, \ldots, n\}$ and $x \ne y$.

Here, X is a group of customers who purchased product x and Y is a group of customers who purchased product y. X, $Y \subseteq U$. So, the similarity between X and Y is defined as

$$J(X, Y) = \frac{|X \cap Y|}{|X \cup Y|} = \frac{|X \cap Y|}{|X| + |Y| - |X \cap Y|} \tag{2}$$

$$0 \le J(X, Y) \le 1$$

Repeat Purchase. In many application of Collaborative Filtering, the common assumption being used is that a customer will buy products that strongly similar to what they already purchased. In the case of offline commerce, this is not only the assumption. We need to capture another customer behavior that tends to buy similar products instead of trying new products they had not bought.

Our approach to deal with this problem is to record any repeat transaction of a certain product up to three times. Thus, we have $3n$ products in total.

$$I = \{i_{11}, i_{12}, i_{13}, i_{21}, i_{22}, i_{23}, \ldots, i_{n3}\} \tag{3}$$

The reason why limits it up to three times transaction is to avoid sparsity problem because not much customers will buy until four times in a given period. The approach will capture the correlation between the purchase of product x for the first time, defined as i_{x1} and purchase of the similar product in the second time, i_{x2}. The more customers bought that product for the second time, the higher probability that the customers had already purchased product x will buy again in the next period. We can see through this simulation:

If the number of customers who purchase product x twice, $|u_{x1} \cup u_{x2}|$, is high thus the Jaccard index, $J(u_{x1}, u_{x2})$, will yield high value as well. This means that there is a strong evidence of repeat purchase of product x. Therefore, we could recommend some customers that already bought product x to buy it for the second time or even third time. To achieve this customer behavior, our customer-product feedback matrix will be like Table 2.

3.2 Item-Based Collaborative Filtering

Item-based Collaborative Filtering is a memory-based approach which produces recommendation based on the relationship between the products inferred from product similarity matrix. The assumption is that a customer will more likely purchase products that are similar or related to the products that he/she has already purchased. Item-based Collaborative Filtering is more efficient than User-based Collaborative Filtering since the model (reduced similarity matrix) is relatively small and can be fully precomputed [4].

Table 2. Adjusted customer-product feedback matrix example.

	i_{11}	i_{12}	i_{13}	i_{21}	i_{22}	i_{23}	...	i_{n3}
u_1	1	1	0	0	0	0	...	0
u_2	1	0	0	1	1	0	...	1
u_3	1	1	1	0	0	0	...	1
...	...	...	...	...	...	...	...	...
u_m	0	0	0	1	1	1	...	1

The aim of Item-based Collaborative Filtering is to generate recommendations for an active customer $u_a \in U$. We define the set of products that are unknown to customer u_a as $I_a = I\backslash\{i_x \in I | r_{ax} = 1\}$. The model building step consists of calculating a similarity matrix S containing all product-to-product similarities using a given similarity measure. In the case of implicit feedback, the most suitable similarity measure is using Jaccard index. Because of containing only value 1 for non-missing, the most popular similarity measures such as Pearson Correlation and Cosine similarity cannot handle this condition [2]. The similarity matrix S is shown in Table 3.

Table 3. Product similarity matrix example.

	i_{11}	i_{12}	i_{13}	i_{21}	...	i_{n3}
i_{11}	–	0.5	0.1	0.3	...	0.1
i_{12}	0.5	–	0.2	0.4	...	0.0
i_{13}	0.1	0.2	–	0.0	...	0.0
i_{21}	0.3	0.4	0.0	–	...	0.1
...	...	...	...	...	...	...
i_{n3}	0.1	0.0	0.0	0.1	...	–

3.3 Top-N Recommendation Product

Once we get the $3n \times 3n$ similarity measure matrix $S = (s_{xy})$, the algorithm then identifies the set of the candidate recommended products by taking the union of the k-most similar product with $k \ll n$. The k products which are most similar to product x is denoted by the set $S(x)$ which can be seen as the neighborhood of size k of the product. This can improve the space and time complexity significantly with the consequence of sacrificing some recommendation quality. Lastly, to make a recommendation based on the model, we use the similarities of other products to calculate a score of the active customer's similarities for related products.

$$\hat{s}_{ay} = \sum_{x \in S(x) \cap \{x | r_{ax} = 1\}} s_{xy} \tag{4}$$

Finally, the candidate products are sorted in decreasing order with respect to that score and the first N products are selected as the Top-N recommended product set.

4 Results and Discussion

4.1 Data Sets

We built the Item-based Collaborative Filtering (IBCF) using an offline commerce's two years transactional data. There are around two million active customers and approximately three hundred different products. It contains customer ID, purchased product and purchase date. Before we go to the model building, the first procedure should be data cleaning especially for reducing data sparsity and synonymy. Initially, we have 316 different products in the environment. We can see its sparsity through the Jaccard index for product-to-product shown in Fig. 1.

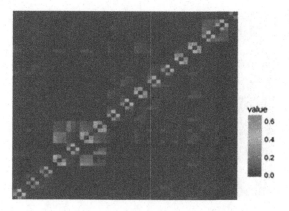

Fig. 1. Sparsity data before manipulation (Color figure online)

The green color means there is a strong correlation between two products, while red color means no correlation that tells us that very few customers purchase those two products. In Fig. 1, the red color is very far dominant compared to the green. This phenomenon is called as sparsity problem in the data. Dimensionality reduction on product side come to overcome this obstacle although it cannot guarantee to vanish all sparse data, it really helps to gain better performance of Collaborative Filtering.

After applying dimensionality reduction using Jaccard index, we eliminate the number of the product to 260 different products. It means we reduced around 17.7% of the original list of products. Our criteria are simple, just looked at the whole Jaccard matrix and see which products that being clustered as a consequence of high similarity. Thus, we can group the products within the cluster as one representative product. As a result, the green color spread more in Fig. 2. In other words, we have successfully reduced the sparsity of data although we cannot remove the whole.

4.2 Experimental Design

We evaluated the performance of the Top-*N* recommendation algorithms in increasing the redeem rate of a personalized promotion comparing to random target selection of most popular products promotion as a control group.

We made the prediction and tested it to customers in six months by dividing into two groups: a group of customers that received personalized promotion and a group of other customers that received most popular products promotion randomly as a control group. Since the limitation of offline commerce marketing channel, we only send content promotion more general by using brand level instead of SKU (Stock Keeping Unit).

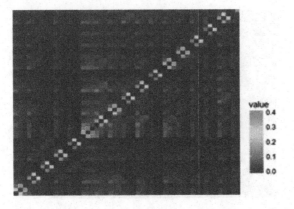

Fig. 2. Sparsity data after manipulation (Color figure online)

We measure how many customers that received the promotion and then go to the store within the given period to buy the offered product as redeem rate,

$$rate_a = \frac{Freq(U_a|r_{ax} = 1)}{|U_a|} \tag{5}$$

where $U_a \in U$ is a set of customers that received personalized offers based on IBCF results. We use the same measurement to control group,

$$rate_b = \frac{Freq(U_b|r_{bx} = 1)}{|U_b|} \tag{6}$$

where $U_b \in U$ is a group of customers that received product recommendations randomly. Note that $U_a \neq U_b$.

4.3   Experimental Results

Our finding is that the redeem rate of personalized promotion is always three times higher over the control group from the first month until the sixth month.

Every month, we sent offers to on average 106,914 customers, for each group of receiving product recommendation based on IBCF output and by random selection. Out of that 641,484 customers in total, 9,738 customers of the IBCF group came to redeem whilst only 3,277 customers of the random group, this gives a redeem ratio between IBCF and random group to about 3:1. Table 4 gives the detail of our experimental results. The variation of total customers being tested is due to the variation of transactions occurred within each month.

From the Table 4 we can make a chart as Fig. 3. It has clearly shown that the recommender systems applied to the offline commerce have extremely helped them increase their marketing return. It had proven through the redeem rate of personalized promotion using IBCF that consistently achieved three times higher than their base products promotion.

Table 4. Redeem rate comparison

Month	Sample size of each group	IBCF group redeem rate	Random group redeem rate
Month 1	204,283	1.4%	0.5%
Month 2	128,416	1.0%	0.3%
Month 3	56,994	1.4%	0.3%
Month 4	151,108	2.3%	0.9%
Month 5	56,234	1.4%	0.4%
Month 6	44,449	1.0%	0.3%

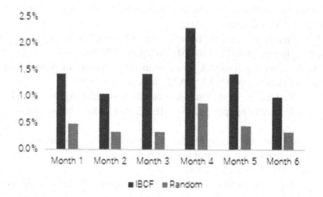

Fig. 3. Redeem rate of IBCF and control group by month

5   Conclusion and Future Work

Item-based collaborative filtering is a suitable algorithm for an offline retail business that does not have rating systems and no sufficient product description. Importantly, IBCF for implicit feedback could be used to successfully capture the customer behavior within the offline commerce that tend to buy a similar product. By building Top-N Item-based Collaborative Filtering, we can infer the products that are most likely be purchased by the customers.

Therefore, it is clear that the product recommender systems could help in increasing the effectiveness of offline commerce marketing and improving their revenue.

One thing that is interesting to be considered for the next project is about the seasonal pattern of customers purchase behavior. This may lead to the use of Model-based algorithm by including time dimension into the customer-product matrix. This could be future work to be implemented within offline commerce recommender systems.

References

1. Hahsler, M.: Developing and testing top-n recommendation algorithms for 0–1 data using recommenderlab. NSF Industry University Cooperative Research Center for Net-Centric Software and System (2011)
2. Hahsler, M.: Recommenderlab: a framework for developing and testing recommendation algorithms. R package version 0.1-5 (2014)
3. Hu, Y., Koren, Y., Volinsky, C.: Collaborative filtering for implicit feedback datasets. In: 8th IEEE International Conference on Data Mining, pp. 263–272. IEEE, Pisa (2008). https://doi.org/10.1109/icdm.2008.22
4. Karypis, G.: Evaluation of item-based top-n recommendation algorithms. In: 10th ACM International Conference on Information and Knowledge Management, pp. 247–254. ACM, Georgia (2001). https://doi.org/10.1145/502585.502627
5. Melville, P., Mooney, R., Nagarajan, R.: Content-boosted collaborative filtering for improved recommendations. In: 18th AAAI National Conference on Artificial Intelligence, pp. 187–192. AAAI, Edmonton (2002)
6. Nichols, D.: Implicit rating and filtering. In: 5th DELOS Workshop on Filtering and Collaborative Filtering, vol. 12. ERCIM, Budapest (1997)
7. Oard, D., Kim, J.: Implicit feedback for recommender systems. In: 15th AAAI Workshop on Recommender Systems, pp. 81–83. AAAI, Wisconsin (1998)
8. Pan, R., Zhou, Y., Cao, B., et al.: One-class collaborative filtering. In: 8th IEEE International Conference on Data Mining, pp. 502–511. IEEE, Pisa (2008). https://doi.org/10.1109/icdm.2008.16
9. Sarwar, B.: Analysis of recommendation algorithms for e-commerce. In: 2nd ACM International Conference on Electronic Commerce, pp. 285–295. ACM, Minnesota (2000). https://doi.org/10.1145/352871.352887
10. Sarwar, B., Karypis, G., Konstan, J., et al.: Item-based collaborative filtering recommendation algorithms. In: 10th ACM International Conference on World Wide Web, pp. 285–295. ACM, Hong Kong (2001). https://doi.org/10.1145/371920.372071
11. Su, X., Khoshgoftaar, T.: A survey of collaborative filtering techniques. Adv. Artif. Intell. **2009**, 19 (2009). https://doi.org/10.1155/2009/421425. Article ID 421425

Malicious Behaviour Analysis on Twitter Through the Lens of User Interest

Bandar Alghamdi[1,2(✉)] ⓘ, Yue Xu[1(✉)] ⓘ, and Jason Watson[1(✉)] ⓘ

[1] Faculty of Science and Engineering, Queensland University of Technology Australia,
Brisbane City 4000, Australia
bandar.alghamdi@hdr.qut.edu.au, {yue.xu,ja.watson}@qut.edu.au
[2] Institute of Public Administration, Riyadh City 11141, Saudi Arabia
alghamdib@ipa.edu.sa

Abstract. Evolving behaviours by spammers on online social networks continue to be a big challenge; this phenomenon has consistently received attention from researchers in terms of how they can be combated. On micro-blogging communities, such as Twitter, spammers intentionally change their behavioural patterns and message contents to avoid detection. Understanding the behavior of spammers is important for developing effective approaches to differentiate spammers from legitimate users. Due to the dynamic and inconsistent behaviour of spammers, the problem should be considered from two different levels to properly understand this type of behaviour and differentiate it from that of legitimate users. The first level pertains to the content, and the second, to the users' demographics. In this paper, we first examine Twitter content relating to a particular topic, extracted from one hashtag, for a dataset comprising both spammers and legitimate users in order to characterise user behaviour with respect to that topic. We then investigate the users' demographic data with a focus on the users' profile description and how it relates to their tweets. The result of this experiment confirms that, in addition to the content level, users' demographic data can present an alternative approach to identify the different behaviours of both spammers and legitimate users; moreover, it can be used to detect spammers.

Keywords: Spam · Spammers · Behaviour · Detection · Text mining
Social networks

1 Introduction

Recent developments in the field of online social networks (OSNs) have led to their integration into nearly all aspects of everyday activity; yet, spammers have taken advantage of those services for malicious purposes. Twitter, for example, is one of the most popular micro-blogging services used widely by spammers to post spam tweets (Grier 2010). Spam tweets refer to unsolicited tweets containing malicious links that direct victims to external sites containing malware downloads, phishing scams, drug sales, etc. (Benevenuto et al. 2010). Researchers have shown an increased interest in fighting web spam (Gyöngyi et al. 2004), email spam (Wang et al. 2013) and online social spammers (Benevenuto et al. 2010; Eshraqi et al. 2015;

© Springer Nature Singapore Pte Ltd. 2018
Y. L. Boo et al. (Eds.): AusDM 2017, CCIS 845, pp. 233–249, 2018.
https://doi.org/10.1007/978-981-13-0292-3_15

Wang et al. 2013; He et al. 2014). According to Stringhini et al. (2010) 83% of users have received at least one unwanted friend request or message on social networks. Moreover, most spam posted on Twitter takes the form of short URLs, and the blacklists used in Twitter were found to be ineffective in the detection of spam (Thomas 2011a b).

Existing approaches used classification based methods (Ma et al. 2009; Ruan et al. 2016; Shen et al. 2017) to detect spammers and malicious URLs in OSNs. Which firstly identify features to represent malicious users or posts, and then classification algorithms are used to classify users or posts into malicious or legitimate. Nnevertheless, the challenge is that these features evolve over time or that they require a significant extraction time and resources. Traditional classification techniques can sort URLs by using host, domain or lexical features (Ma 2009; Kotsiantis et al. 2006; Feroz and Mengel 2014; Thomas 2011a, b). However, these techniques do not work well on OSNs because OSNs are vast and dynamic; the content changes every moment as posts are added, shared and updated. In addition, with short URLs, especially those used on Twitter, it becomes challenging to tell spam URLs apart from legitimate ones as short URLs change the look of the URL and hide the real domain name. Thus, there is a need to find new features that can identify spammers and malicious URLs in OSNs without solely relying on the existing feature selection techniques.

Alternatively, understanding the behaviour of spammers when they connect to social networking sites creates better opportunities for extracting features and improves the detection of malicious behaviour. Some recent additions to the literature have offered valuable findings about spammers' behaviour (Martinez-Romo and Araujo 2013; Cao and Caverlee 2014; Ruan et al. 2016; Dang et al. 2016). These sophisticated approaches employ supervised machine learning techniques using features such as: usage of hashtags and retweeting behaviour to identify spammer's behaviour more accurately than traditional approaches. However, a major problem with these approaches is that they investigate users' behaviour based on incomplete view of user-generated information. Due to the dynamic and inconsistent behaviour of spammers, the current approaches failed to combat these issues properly (Shen et al. 2017). Thus, there is a need to find new features that can identify malicious behaviour and URLs in OSNs. This would allow OSNs to be viewed from more than one angle, especially from the users' demographic, profile description and generated content.

In order to bridge the gaps expressed through the aforementioned challenges and to detect malicious behaviour effectively, this paper proposes an approach for analysing the behaviour of spammers where the relevant features of users' demographic data are viewed together with content generated from the perspective of user interests (Alghamdi et al. 2016). Since attackers intend to propagate malicious URLs, their interests change frequently so they can exploit any event that is trending or that has active users. So far, to our knowledge, this method has not yet been applied to spam behaviour analysis in online social networks, and this is the first work to do so. We first analyse tweet content pertaining to one topic—extracted from one hashtag—for both spammers and legitimate users, using Frequent Patterns (FP) (Agrawal 1993) and Term Frequency (TF) to characterise the behaviour in relation to that topic. By analysing the tweet content pertaining to one topic, we observed that spammers use different words than legitimate users on a given topic as they are semantically disconnected from the topic. This observation then led us to investigate users' demographic data as there is a relationship between an individual's posts and his or her motivations—namely, interests—in participating in the topic. The investigation of users' demographic data focuses on the users' profile description and how it relates to the tweets.

In summary, our work differs from the existing literature in the following respects:

- We clearly differentiated between spammers and non-spammers for one topic by applying TF and FP to the tweet content. This technique separates user features and behaviour into a set of distinct patterns.
- We proposed a new approach to identify spammer's behaviour based on users' demographic data, by focusing on users' profile description and how it relates to their tweets. This approach is used to determine the similarity between the user's profile descriptions and tweets based on cosine similarity because spammers tend to show either lower similarity or high divergence between their profile description and their tweets.

The experimental results obtained using a real-world Twitter dataset showed that the proposed features of users' content and demographic data—especially user interests and time activities—can present an alternative approach to observe different behaviours of both spammers and legitimate users. This approach can also be used to detect malicious users.

2   Related Work

A considerable amount of literature has been published on spam detection in social sites and researchers have shown an increased interest in this regard. (Heymann et al. 2007) firstly surveyed potential solutions and challenges on social spammer detection, many different methods have been proposed to combating social spammers.

There are some work train classifiers using **Content-based** features. (Bhattarai et al. 2009) used different features to detect spam comment in a blog that targets short text. Features, such as post-comment similarity, word duplication, concentration of noun phrases and stop words ratio, were used to detect spam text. These features would be insufficient in OSNs unless combined with others because using content features alone is not sufficient as spammers have become very smart and their behaviour includes inserting some legitimate text within the malicious text. (Cao and Caverlee 2014) Proposed a model that is more robust in identifying the malicious URLs that are being shared using behavioral examination of URL sharing from two different perspectives: how these links are poste and received. However, the main limitation of this study is that it only considered one type of URL -Bitly URL- because it provided statistical click information. Another study (He et al. 2014) pointed out that the behaviour of users on a social network, particularly on Twitter, depends on many different factors, such as interests, hobbies, time spent. They used the time complexity of posting behaviour, and multi-scale entropy method to measure the complexity of posting behaviour. The result suggested different types of users on Twitter such as: individual accounts, news platform accounts, advertising accounts and robot accounts (Dang et al. 2016).

The use of Frequent-Patten (FP) to predict malicious URL was introduced by (Zheng et al. 2015). They firstly, found relationship between URLs through the relationship between people and URLs. Then the association rules between malicious URLs and people were mined on the base of events of individuals and URLs. Finally, various communities by modularity can be constructed where each community represents different types of URLs. (Shen et al. 2017) used bag-of-words and took term frequency-inverse document frequency (tf-idf) for tweet content as feature value (text distributions). They also considered the frequency of URL and hashtag posted by a particular user to extract user's features and take social relationships into consideration. Nevertheless, the strategy did not take the valuable information presented in the user's profile into consideration. In addition, building social relationships requires a significant extraction time.

Some works only consider user's **profile** to identify spammers without considering content features. Profile-Pattern is considered to be a fast way to detect malicious and spam accounts without analysing tweets - content. (Thomas 2011a, b) used pattern-recognition to detect fake profiles through the connections between profile information, such as email, screen name and profile name. Although the model they proposed for detection performed well, their study was limited to market accounts that only sold special products. (Benevenuto et al. 2010) captured user's behaviour through profiles. They considered 23 profiles attributes, such as the number of tweets, number of followers, number of followees, the age of the account and the number of mentioned replies. Due to the large number of features and some irrelevant features, they classified only 70% of the spammer accounts correctly. (Hua and Zhang 2013) noted, particularly for Twitter, that spam and non-spam profiles overlap, which can make it a challenge to identify malicious content across a network. However, certain characteristics are prominent among spam profiles, including a young account, tweets with a higher succession rate, tweets with greater status, tweets that contain spam words.

A part from user's profile is **time activities** between different posts. (Gao et al. 2010) proposed the FBCluster Scheme to detect spam campaigns on Facebook through the temporal characteristics between first and last spam post of the user. They relied heavily on the busty and distributed nation of spam messages, and mistakenly labelled non-spam post as spam with 39.3% false positive rate FPR. Several studies discussed user profile-based and content-based features together to provide a full understanding of the behaviour of spammers (Nepali and Wang 2016; Shen et al. 2017; Jeong et al. 2016; Mccord and Chuah 2011; Yang et al. 2011). They determined that there was a strong and consistent correlation between the profile and the content for all suspicious accounts. This present study extended this to examine user features from multiple views, including messages, URLs and hashtags. By including user interest information, we aim to develop a new process for detecting malicious users with high reliability.

3   Data Collection

The dataset, American Airlines Twitter Dataset - AAL, which we used in the investigation, was obtained from NASDQ[1] an online repository that provides free Twitter datasets. This dataset was collected from 28 March 2016 to 15 June 2016, and the information was about American Airlines Group $ALL. This dataset was collated using all tweets mentioning the company name, or Twitter Symbol ($ALL). The dataset contained 6509 tweets. Each tweet has attributes, such as tweet text, retweets, hashtags, URLs, etc. The dataset also contained some user profile information, such as screen name, user name, description, number of followers and number of people the user is following. Table 1 lists the general statistics of the dataset. Because the main goal of this investigation is to study the differences between spammers and legitimate users, we labelled each tweet as malicious or legitimate (non-malicious) using some publically-available blacklists based on the URLs appearing in the tweet. Therefore, we extracted tweets that have valid URLs and we omitted tweets that did not have valid URLs. All the URLs were validated by using Python Library requests and *response*.

Table 1. General statistics of dataset.

	Number of tweets	Number of users
All tweets	6509	1657
Tweets with URL	4800	850
Malicious tweets	919	44
Legitimate tweets	3881	740

Table 2. Top frequent hashtags for malicious and non-malicious tweets.

Non-malicious		Malicious	
Hashtag	Relative support	Hashtag	Relative support
#AAL	0.135	#AmericanAirlines	0.476
#daytrading	0.084	#AngloAmerican	0.366
#stock	0.069	#Fortune50	0.13
#stocks	0.069	#APAC	0.02
#investing	0.058	#ASX2000	0.02
#Stocks	0.054	ASX2000, #APAC	0.02
#finance	0.043		
#invest	0.042		
#stockmarket	0.038		
#pennystocks	0.037		

---

[1] http://www.followthehashtag.com/datasets/nasdaq-100-companies-free-twitter-dataset/.

3.1 Labelling URLs

Each URL was labelled based on two blacklist lookup services: VirusTotal[2] and Web of Trust[3]. For each of the blacklists, we posted the URL using the API for each service to obtain the response result, and this technique has been used in different research studies (Dewan and Ponnurangam 2015) (Martinez-Romo and Araujo 2013). The response from VirusTotal has multiple domain information, such as Google Safe Browsing, PhishTank, Dr.Web, etc., with a total of 68 domains. For our data, we considered the URL to be malicious if at least one domain marked it as malicious. This is because each domain has its own way of detecting a malicious URL. For example, PhishTank uses the lexical feature of the URL and domain to label it, whereas other services use the IP address. This makes VirusTotal a reliable source for labelling our dataset.

The second way of labelling a URL is using the Web of Trust API that calculates the reputation of a given domain. The trustworthiness of the domain is measured in pairs {domain, component} and computed as two values: a reputation estimate and confidence in the reputation for each given component. Any value less than 60 is considered to be unsatisfactory, and we marked the URL as malicious if this condition was satisfied. In addition to reputation, Web of Trust also categorises a website based on third party and user votes. The system has different categories, and we marked the URL as malicious if the following categories were found in the response of the Web of Trust API: malware or viruses, phishing, scam, potentially illegal, privacy risk, spam and potentially unwanted program[4].

4 Content Comparison

4.1 Hashtags Distribution

The aim is to investigate the differences between malicious users and legitimate (non-malicious) users in terms of their tweet content; Frequent Term (FT) is a method that can help achieve this task. Frequent terms are usually used as features to represent the content of a text document. Therefore, frequent terms can be used to represent the content of the users' tweets, from which we can compare the content differences between malicious users and non-malicious users at the content level. Twitter users can use hashtags in a tweet to indicate the topics addressed in the tweet. A number of studies investigated Twitter hashtags to understand malicious behaviour. (Sedhai and Sun 2017) (Shen et al. 2017)found that hashtags can provide information at the tweet-level and they can be used to identify spammer behaviour. In this section, we will first look at the distribution of malicious tweets among hashtags in the entire dataset. We grouped the tweets based on the hashtags for malicious and non-malicious tweets to understand a specific aspect of user behaviour.

---

[2] https://www.virustotal.com/.
[3] https://www.mywot.com/.
[4] https://www.mywot.com/wiki/API.

It is important to mention that our dataset has one main topic, which is the American Airlines Group[5]. However, we assume that the dataset has multiple sub-topics (hashtags) that indicate more specific topics. The hashtags can help in understanding the behaviour of malicious users in regard to a specific topic. Initially, we have to represent all the hashtags for the entire dataset, and then we have to observe where the malicious tweets take place among these hashtags. Figure 1 shows the top 32 frequent hashtags and the distribution of tweets among each hashtag; the black colour represents all the tweets on each hashtag, the red colour represents the malicious tweets for the same hashtag and the green colour represents the legitimate (non-malicious) tweets.

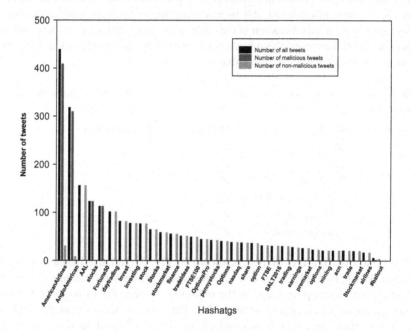

Fig. 1. Top 32 hashtags and the distribution of malicious tweets among hashtags. (Color figure online)

Out of 359 hashtags, 271 have less than six tweets, which represent 75% of the hashtags; only 2.2% of the hashtags have over 100 tweets. In general, 97.7% of the all the hashtags have less than 100 tweets, as shown in Table 3. We generated a frequency pattern for the hashtags in the dataset. The top frequent hashtags in the malicious and non-malicious tweets are listed in Table 2.

---

[5] http://www.followthehashtag.com/datasets/nasdaq-100-companies-free-twitter-dataset/.

Table 3. Hashtag-based statistics and tweets.

Hashtag	Relative support
157 hashtags	One tweet in each hashtag
271 hashtags	Less than six tweets in each hashtag
351 hashtags	Less than 100 tweets in each hashtag
8 hashtags	Over 100 tweets in each hashtag

For the hashtags that have malicious tweets, four were found to be completely malicious for example #Fortune50. All tweets using those hashtags were malicious and another two were a mix between some legitimate (non-malicious) tweets and some malicious tweets; however, these hashtags were still deemed malicious as most of the tweets in them were malicious see Table 4.

Table 4. Hashtag with malicious tweets and hashtags with mixture of malicious and non-malicious.

Hashtag	Malicious	Non-malicious
#Fortune50	✓	
#APAC	✓	
#ASX2000	✓	
#ASX2000, #APAC	✓	
#AmericanAirlines	✓	✓
#AngloAmerican	✓	✓

From the distributions of tweets shown in Fig. 1, and the frequent pattern of hashtags shown in Table 3 we can see that some of the top hashtags in the entire dataset are also the top hashtags in the malicious tweets; for example, #American-Airlines, AngloAmerican and #Fortune50. This clearly shows that the malicious users target these trending topics, which is also confirmed in the finding from previous studies(Martinez-Romo and Araujo 2013) (Mccord and Chuah 2011). We also observed that a number of hashtags occurred only in malicious tweets, such as #Fortune50 and #APAC. Thus, these hashtags were never used by legitimate (non-malicious) users. A malicious user or a group of malicious users might add these hashtags. These types of hashtags have been investigated by (Dang et al. 2016) who found that a group of spammers tended to work cooperatively to add anomalous topics and use retweet features as a group to spread malicious content as much as possible. In our dataset, it was obvious that there was not much retweeting behaviour for these hashtags, and this is an indication of evolving spammer behaviour. This could be due to a group of spammers having different accounts and posting to the same hashtag, or it might be due to multiple accounts managed by one spammer using Twitter application.

4.2   Content Comparison Under a Specific Topic

The distribution shown in Fig. 1, and the variation pattern in the hashtags shown before, can be attributed to differences in the user's engagement with a particular topic. This led us to further investigate the user-level to uncover the implicit behaviour that explains this diversity. However, before doing so, we compared the content of the tweets for both spammers and normal users for one hashtag (one sub-topic). Applying the Frequent-Pattern method will help us to represent each hashtag group. To the best of our knowledge, this study was the first study to use FP in Twitter content for malicious detection. We assumed that the FP for spammers and legitimate tweets for one topic had different representations. Therefore, we generated FPs of hashtag AmericanAirlines and AngloAmerican. The results are shown in Figs. 2 and 3. Both figures show that there are differences in patterns for spammers and legitimate users under one topic, and many terms have a relative support count of 0 in the malicious tweets, but the same terms have a higher value in the legitimate tweets. This shows an overfitting in both cases. For example, for sub-topic 1, the term 'aal', which is the term and hashtag that dominated the dataset, occurred frequently in legitimate tweets, but it did not occur in the spam tweets. This difference is the outcome of the engagement of malicious users in the discussion of the topic using different words without being semantically involved in the topic, and the pattern shows that they can be distinguished from nonmalicious users. Another example is the frequent occurrence of the term 'read' in the malicious tweets; this term does not occur at all in the non-malicious tweets. (Nepali and Wang 2016) confirmed this observation, and they used it as a feature of the tweet content to classify suspicious short URLs posted on Twitter based on the visible tweet attributes. They found that terms, such as *'check and click'*, exist in malicious tweets.

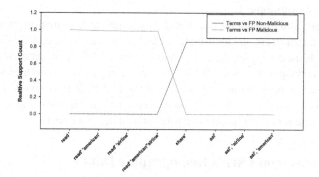

Fig. 2. FP shows the differences between spammers and normal tweets for sub-topic 1 (#AmericanAirlines)

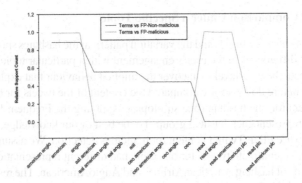

Fig. 3. FP shows the differences between spammers and normal tweets for sub-topic2 (#AngloAmerican)

Similar findings were observed for sub-topic 2, as some similar and opposite patterns were seen. The patterns, 'american anglo', 'american' and 'anglo', are common in both groups, but the patterns of 12 terms in sub-topic 2 are opposite of the that presented in the legitimate tweets, and their support count frequency values are either a high or 0. For example, 'aal american', 'aal' and 'ceo american' are not frequent terms in the malicious tweets, but the same terms have a frequency that ranges from 0.44 to 0.55 in the non-malicious tweets. Moreover, the term 'read' has a high frequency in the malicious tweets of sup-topic 2, and this time they are in conjunction with other terms, such as 'read anglo' and 'read american', which do not exist in the non-malicious tweets. Again, the term 'read' is very common in the malicious tweets, as mentioned in the discussion about sub-topic 1.

In summary, the FP method is a useful way to obtain information about the differences between spammers and legitimate content. It can also be used to identify the most common words used by spammers that are not often used by legitimate (non-malicious) users. The conclusion of using FP for a content comparison of both groups is that a relationship exists between an individual's posts and his/her motivation (interest) to participate in the topic. The next section will examine the user-level analysis, in order to understand the relationship between user interest and user posts.

5   Comparison on User's Demographic Data

5.1 User's Description

Users provide their interest in the description statement contained in their profile. The users' descriptions provide explicit information about their interests, whereas the users' tweets reveal their interest implicitly. As the aim of this study is to understand malicious behaviour through explicit and implicit behaviour, the comparison between a user's description and his/her tweets can help in understanding their behaviour. Therefore, the user can be assessed by analyzing the user's description in relation to the tweets they make. Our aim is to determine if a user description shows any consistency with tweets that are posted by the user. A comparison between a user's description and his/her tweets could provide insights into features that can be used to differentiate malicious users from non-malicious users. We would assume that the inconsistency between the user description and the tweets would be a feature to differentiate.

The description in a user's profile in Twitter is called a Bio. Twitter bio gives the users up 160 text characters to tell the others about themselves, plus, 30 bonus characters for location as well as an opportunity to give a backlink to your own site. This field contains a few simple sentences describing the user's interest to give people the first impression about a user's interest, and many attackers use it to attract more followers. All the studies reviewed so far, however, failed to take the content of Bio into account, which we are seeking to cover in this research project.

In the dataset, we found that 87% of the spammers have a user description. This finding is different from what was reported by Sedhai and Sun (2017), they stated that most spammers tend to have no user description in their user profile. In contrast to (Hua and Zhang 2013), however, particularly for Twitter, that spam and non-spam profiles has no distinctions especially description, location and URL in the profile. This is an indication that spammers tend to have a profile that is look similar to legitimate users. Although, the dataset is relatively small at this stage, it will be changed in the future work to include a massive number of users (millions of users) in order to provide more accurate statistics. For the non-malicious users, dataset contains 740 users who posted over 3881 tweets, and 603 of those users have a user description in their user profile.

The general assumption about the user's description is that it is normal for a user to provide a description that does not match every single tweet he/she posts. For example, if a user writes a description, such as "specialist in children with special needs", this user should have a large number of tweets that match the description, but every single tweet will not necessarily match that description.

While investigating the user's description with the corresponding tweets for all malicious users, we have found that a group of users posted exactly the same massage and the same URL. While this could be considered to be normal behaviour, according to Twitter Reporting Spam[6] repeatedly posting a duplicate message is spam. This was obvious when we looked at their descriptions and found that all of them are the same with only one or two different words, and all of the users included the same hashtag in their description. Additionally, they tweeted from different geographical locations. This could open up a new direction in our research, imploring us to investigate the cooperative attackers in a social network using the relationship between user's description and tweets.

In this regard, a recent study by (Dang et al. 2016) employed the topology-based method and anomalous topic sequences to identify malicious topic using subgraph ranking through retweeting behaviour, and then they labelled the spammers based on the total retweeting that occurred for one topic. However, their method will be useless in our dataset as spammers do not use the retweet option; rather, they use multiple accounts to post the same message.

---

[6] https://support.twitter.com/articles/64986.

5.2   Interest Consistency Between the Description and Tweet Content

Users describe themselves in their profiles, and their behaviour is evident in their tweets. This section addresses explicit and implicit behaviour by analysing user's descriptions and tweets.

At some level, the content of a user's tweets should be consistent with the user's description. We conducted an experiment to evaluate the consistency between the users' description and the content of their tweets for both spammers and normal users. Cosine similarity can be used to measure the similarity between a user's description and the tweet content represented as vectors. Given two vectors A and B the cosine similarity is calculated as follows:

$$Sim(A, B) = \cos(\theta) = \frac{A * B}{\|A\| \|B\|} \tag{1}$$

We generated a vector for each single description of each user, and then we obtained the most frequent words from each user's tweets and selected the top 20 frequent words to create vectors for each user's tweets. This generated a list of vectors that represent each user's description, and we have a list of the users' vectors that represent each user's tweets (each user is represented by one vector). We used tf-idf values to produce these vectors. We then calculated the similarities between each single user's description and all the users' tweets and got the average values.

However, the result was not useful to compare because the non-malicious users have an average of similarity of 0.009905 and the malicious users have an average similarity of 0.00582925, which is very low. Consequently, we measured the cosine similarities between the user's description and every single tweet for both spammers and normal users. In addition to the top 20 words of the user's tweets, we used the entire tweet (each tweet as a vector), and then we calculated the similarity between the user's description and each single tweet for each user, only focusing on the maximum values. The reason for using the maximum value is that it implies that as long as there are some tweets that are similar to the description, the similarity is high; thus, normal (non-malicious) users are relatively consistent with their descriptions. In other words, the similarity values between the user description and the tweets should be higher for non-malicious users than they are for malicious users.

The result was comparatively useful as the non-malicious users had a maximum similarity value ranging between 0.558758 and 0.144609; in contrast, the maximum similarity value for the spammers ranged between 0.313629 and 0. Although, there is overlap between the two groups, it is relatively small. The maximum similarity value for the non-malicious users is over 0.50, indicating higher similarity when we compare it with the top similarity value for the malicious users in which is 0. Figure 4 shows that, overall, the malicious users have lower similarity values between their descriptions and tweets. Moreover, we can see the overlap between them; it is a small portion, and a threshold can be identified in order to use it as a feature to distinguish between these two groups.

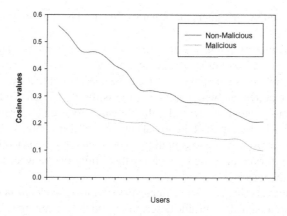

Fig. 4. Cosine similarity, (max value) between tweets and users descriptions.

It would be worth to compare our approach in this section with previous methods mentioned in literature for users' features extraction. (Shen et al. 2017) statistically analyzed the distribution differences between spammers and legitimate users from three views text, URL and hashtag information to represent users' features. However, this analysis does not take account of user's profile information and was limited to content level. Such expositions are unsatisfactory because they are still one single view and cannot represent the use's features properly. Our existing approach, however, considered tweets content: hashtags and URLs with inclusion of user's description content. Ignoring the explicit information of the users on the profile for users' features might deliver incomplete picture of users' behavior.

Presence of users' description in the profile was used as a feature to identify non-spammers users by (Sedhai and Sun 2017; Aggarwal et al. 2012) yet no actual analysis of the text of the user's description was done. The aforementioned researchers would have been more relevant features if user's description has been explored. Consequently, it has been considered in this study and had usefully analysed to identify different behavior patterns of spammers and non-spammers users.

6 Discussion

The strong relationship between user's features, URL features and OSN features has been described in the literature. Previous studies have emphasized the importance of URL features and OSN features in detecting spammers or malicious URLs; however, most of these previous works investigated these features separately, with only few studies considering them together. The present study aimed to integrate these features into a framework from the perspective of user interests in order to increase the level of detection and provide more reliable features that cannot be easily evaded by spammers.

The recommended approach in this paper involved the analysis of FP and TF to enable additional inferences about the content. FP provided implicit information about the differences between spam content and normal content; it also revealed the most common words used by spammers that are not often used by legitimate (non-malicious) users.

We extracted more reliable features from users' demographic data. The current features in the existing literature, such as the number of followers, the number of users being followed and the use of spam words, can be easily modified by an attacker. This modification of features enables spammers to pass through Twitter's filtering system and post malicious content. Conversely, this research focused on features that do not evolve over time, which are part of the social network infrastructure. The proposed work achieved improved results from information on user interests extracted from the profile description; this information is used to measure the similarities between the users' profile description and the tweets.

The experiment conducted on user interests took into consideration the fact that not every single tweet matched the profile description, and the maximum value of the cosine similarity provided comparative values for malicious and non-malicious users. It is worth mentioning that the framework and the experiment conducted in this study to detect spammers presented the following strengths:

1. This work considered user interests since the engagement of users in any activity is driven by their interests; further, this feature is difficult for a spammer to manipulate given that the behaviour of spammers lacks a focused interest.
2. The URLs in the posted content can be uncovered by user behaviour from the tweet content using FP and based on the user's demographic data; the entire page need not be downloaded for analysis.
3. The framework can handle incomplete information about the user given that spammers tend to set up an account using only their first and second name and an email address. However, this work can identify this behaviour implicitly through tweet content.

Finally, a number of limitations need to be considered. First, modelling for both posts and profiles entails a significant amount of work. In addition, the current analysis of demographic data, especially with respect to user interests, was limited to the current dataset; however, access to users' historical tweet data will provide a more in-depth understanding of user interests and behaviour. Finally, this research project disregarded the community that users belong to, such as their friends circle; this is relevant because such communities can impact user behaviour and generate valuable inferences about this behaviour.

7   Conclusion

The purpose of this paper was to examine new characteristics of Twitter users and user-generated content in the field of spammer detection. The analysis revealed several features and characteristics that can help distinguish the behaviour of spammers from that of legitimate users. Demographic data and content of users together enabled an alternative approach to observe different behaviours. A content-level investigation of FP for a single topic revealed that spammers are semantically disengaged with the topic.

We also measured the similarity between users' profile descriptions and tweets to identify users with focused interests. The experimental results obtained using real-world data collected from Twitter showed that our proposed features can effectively tell spammers apart from normal, legitimate users.

This work introduces our preliminary effort to identify holistic features in detecting the behaviour of spammers in the context of micro-blogging. As the behaviour of spammers evolves over time, there will be a need for more features that are not dynamic. Future research will need to move in two directions: one, refining our approach with a larger dataset and, two, developing a classifier to detect spammers and test our features.

References

Aggarwal, A., Rajadesingan, A., Kumaraguru, P.: PhishAri: automatic realtime phishing detection on Twitter. In: IEEE eCrime Researchers Summit, Las Croabas, Puerto Rico, p. 1 (2012). https://doi.org/10.1109/eCrime.2012.6489521

Agrawal, R., Imieliński, T., Swami, A.: Mining association rules between sets of items in large databases. In: Proceedings of the ACM SIGMOD International Conference on Management of Data, Washington, D.C., USA, pp. 207–216 (1993)

Alghamdi, B., Watson, J., Xu, Y.: Toward detecting malicious links in online social networks through user behavior. In: IEEE WIC/ACM International Conference on Web Intelligence Workshops, Omaha, USA, pp. 5–8 (2016). https://doi.org/10.1109/wiw.2016.014

Benevenuto, F., Magno G., Rodrigues, T., Almeida, V.: Detecting spammers on Twitter. In: Collaboration, Electronic Messaging, Anti-abuse and Spam Conference (CEAS), vol. 6, p. 12 (2010)

Bhattarai, A., Rus, V., Dasgupta, D.: Characterizing comment spam in the blogosphere through content analysis. In: IEEE Symposium on Computational Intelligence in Cyber Security, CICS 2009, pp. 37–44 (2009)

Cao, C., Caverlee, J.: Behavioral detection of spam URL sharing: posting patterns versus click patterns. In: International Conference on Advances in Social Networks Analysis and Mining (ASONAM), pp. 138–141. IEEE/ACM, Beijing (2014)

Dang, Q., Zhou, Y., Gao, F., Sun, Q.: Detecting cooperative and organized spammer groups in micro-blogging community. Data Min. Knowl. Discov. **31**, 573–605 (2016)

Dewan, P., Ponnurangam, K.: Towards automatic real time identification of malicious posts on Facebook. In: 13th Annual Conference on Privacy, Security and Trust (PST), Izmir, pp. 85–92 (2015)

Eshraqi, N., Jalali, M., Moatar, H.M.: Detecting spam tweets in Twitter using a data stream clustering algorithm. In: International Congress on Technology, Communication and Knowledge (ICTCK), pp. 347–351 (2015)

Feroz, M.N., Mengel, S.: Examination of data, rule generation and detection of phishing URLs using online logistic regression. In: International Conference on Big Data, pp. 241–250. IEEE, Washington, D.C. (2014)

Gao, H., Hu, J., Wilson, C., Li, Z., Chen, Y., Zhao, B.Y.: Detecting and characterizing social spam campaigns. In: 10th ACM SIGCOMM Conference on Internet Measurement, pp. 35–47. ACM (2010)

Grier, C., Paxson, V., Zhang, M.: @spam: the underground on 140 characters or less. In: ACM Conference on Computer and Communications Security, pp. 27–37. ACM (2010)

Gyöngyi, Z., Garcia-Molina, H., Pedersen, J.: Combating web spam with trustrank. In: Thirtieth International Conference on Very large Data Dases, vol. 30, pp. 576–587. VLDB Endowment (2004)

He, S., Wang, H.J., Hong, Z.: Identifying user behavior on Twitter based on multi-scale entropy. In: International Conference on Security, Pattern Analysis, and Cybernetics (SPAC), pp. 381–384. IEEE, Wuhan (2014)

Heymann, P., Koutrika, G., Garcia-Molina, H.: Fighting spam on social web sites: a survey of approaches and future challenges. IEEE Internet Comput. **11**(6), 36–45 (2007)

Hua, W., Zhang, Y.: Threshold and associative based classification for social spam profile detection on Twitter. In: The Ninth International Conference on Semantics, Knowledge and Grids (SKG), pp. 113–120. IEEE, Beijing (2013)

Jeong, S.Y., Koh, Y.S., Dobbie, G.: Phishing detection on Twitter streams. In: Cao, H., Li, J., Wang, R. (eds.) PAKDD 2016. LNCS (LNAI), vol. 9794, pp. 141–153. Springer, Cham (2016). https://doi.org/10.1007/978-3-319-42996-0_12

Kotsiantis, S.B., Zaharakis, I.D., Pintelas, P.E.: Machine learning: a review of classification and combining techniques. Artif. Intell. Rev. **26**(3), 159–190 (2006)

Ma, J.S., Savage, L.K., Voelker, S., Geoffrey, M.: Beyond blacklists: learning to detect malicious web sites from suspicious URLs. In: ACM SIGKDD International Conference on Knowledge Discovery and Data Mining, pp. 1245–1254. ACM, New York (2009)

Martinez-Romo, J., Araujo, L.: Detecting malicious tweets in trending topics using a statistical analysis of language. Expert Syst. Appl. **40**(8), 2992–3000 (2013)

McCord, M., Chuah, M.: Spam detection on Twitter using traditional classifiers. In: Calero, J.M.A., Yang, L.T., Mármol, F.G., García Villalba, L.J., Li, A.X., Wang, Y. (eds.) ATC 2011. LNCS, vol. 6906, pp. 175–186. Springer, Heidelberg (2011). https://doi.org/10.1007/978-3-642-23496-5_13

Nepali, R.K., Wang, Y.: You look suspicious!!: leveraging visible attributes to classify malicious short URLs on Twitter. In: Hawaii International Conference on System Sciences (HICSS), pp. 2648–2655. IEEE (2016)

Ruan, X., Wu, Z., Wang, H., Jajodia, S.: Profiling online social behaviors for compromised account detection. IEEE Trans. Inf. Forensics Secur. **11**(1), 176–187 (2016). https://doi.org/10.1109/TIFS.2015.2482465

Sedhai, S., Sun, A.: An analysis of 14 Million tweets on hashtag-oriented spamming. J. Assoc. Inf. Sci. Technol. **68**(7), 1638–1651 (2017)

Shen, H., Ma, F., Zhang, X., Zong, L., Liu, X., Liang, W.: Discovering social spammers from multiple views. Neurocomputing **225**, 49–57 (2017)

Stringhini, G., Kruegel, C., Vigna, G.: Detecting spammers on social networks. In: Proceedings of the ACM of the 26th Annual Computer Security Applications Conference (ACSAC 2010), pp. 1–9. ACM, New York (2010)

Thomas, K., Grier, C., Ma, J., Paxson, V., Song, D.: Design and evaluation of a real-time URL spam filtering service. In: IEEE Symposium on Security and Privacy (SP), Berkeley, CA, pp. 447–462. IEEE (2011a)

Thomas, K., Grier, C., Song, D., Paxson, V.: Suspended accounts in retrospect: an analysis of Twitter spam. In: The 2011 ACM SIGCOMM Conference on Internet Measurement conference, New York, USA, pp. 243–258. ACM (2011b)

Wang, D., Irani, D., Pu, C.: A study on evolution of email spam over fifteen years. In: 9th IEEE International Conference on Collaborative Computing: Networking, Applications and Worksharing, (Collaboratecom), Austin, TX, pp. 1–10. (2013)

Wang, D., Navathe, S.B., Liu, L., Irani, D., Tamersoy, A., Pu, C.: Click traffic analysis of short URL spam on Twitter. In: International Conference Conference on Collaborative Computing: Networking, Applications and Worksharing (Collaboratecom), IEEE, pp. 250–259 (2013)

Yang, C., Harkreader, R.C., Gu, G.: Die free or live hard? Empirical evaluation and new design for fighting evolving Twitter spammers. In: Sommer, R., Balzarotti, D., Maier, G. (eds.) RAID 2011. LNCS, vol. 6961, pp. 318–337. Springer, Heidelberg (2011). https://doi.org/10.1007/978-3-642-23644-0_17

Zheng, L.-X., Xu, X.-L., Li, J., Zhang, L., Pan, X.-C., Ma, Z.-Y., Zhang, L.-H.: Malicious URL prediction based on community detection. In: 2015 International Conference on Cyber Security of Smart Cities, Industrial Control System and Communications (SSIC), pp. 1–7. IEEE (2015)

Meta-Heuristic Multi-objective Community Detection Based on Users' Attributes

Alireza Moayedekia[1(✉)], Kok-Leong Ong[2], Yee Ling Boo[3],
and William Yeoh[1]

[1] Department of Information Systems and Business Analytics,
Deakin University, Geelong, VIC 3125, Australia
{amoayedi,william.yeoh}@deakin.edu.au
[2] SAS Analytics Innovation Lab, ASSC, La Trobe University,
Melbourne, VIC 3086, Australia
kok-leong.ong@latrobe.edu.au
[3] School of Business IT & Logistics, RMIT University,
Melbourne, VIC 3000, Australia
yeeling.boo@rmit.edu.au

Abstract. Community detection (CD) is the act of grouping similar objects. This has applications in social networks. The conventional CD algorithms focus on finding communities from one single perspective (objective) such as structure. However, reliance on only one objective of structure. This makes the algorithm biased, in the sense that objects are well separated in terms of structure, while weakly separated in terms of other objective function (e.g., attribute). To overcome this issue, novel multi-objective community detection algorithms focus on two objective functions, and try to find a proper balance between these two objective functions. In this paper we use Harmony Search (HS) algorithm and integrate it with Pareto Envelope-Based Selection Algorithm 2 (PESA-II) algorithm to introduce a new multi-objective harmony search based community detection algorithm. The integration of PESA-II and HS helps to identify those non-dominated individuals, and using that individuals during improvisation steps new harmony vectors will be generated. In this paper we experimentally show the performance of the proposed algorithm and compare it against two other multi-objective evolutionary based community detection algorithms, in terms of structure (modularity) and attribute (homogeneity). The experimental results indicate that the proposed algorithm is outperforming or showing comparable performances.

Keywords: Attributed communities · Community detection · Harmony search

1 Introduction

Community detection (CD) is grouping objects with similar characteristics [28]. Conventional solutions to community detection rely on one single objective function. The shortcoming of reliance on single objective function (e.g., structure properties) is that it does not help in identifying nodes that are well-separated, while densely intra-connected [3, 14, 20]. As a result of this conventional algorithms to CD will

© Springer Nature Singapore Pte Ltd. 2018
Y. L. Boo et al. (Eds.): AusDM 2017, CCIS 845, pp. 250–264, 2018.
https://doi.org/10.1007/978-981-13-0292-3_16

become biased on the obtained community partitions, and they may not be able to detect multiple potential structures [6].

To overcome this issue the state-of-the-art CD algorithms either rely on some heuristics to derive network partitions by executing some heuristic rules, or formulating community detection as a combinatorial optimization problem and detecting the community structure by optimizing a predefined evaluation criterion which describes a certain property of community, such as modularity, normalized cut and the map equation [9]. Hence, CD algorithms are formulated as Multi-objective Optimization Problems (MOPs), that describe multiple properties of networks by optimizing multiple conflicting (but complementary) criteria and obtain multiple network partitions which correspond to different trade-offs amongst these criteria [18, 19].

By overcoming this issue, state-of-the-arts can take full advantage of attribute information to partition the networks intuitively [10, 16]. They also can take the nodes' attribute information as input and group objects with the same attribute values into the same groups. This is done through either pattern mining [12, 15] or network partition [27] methods. Pattern mining methods lack flexibility as they force every node in each community to have same or similar attribute values. This may lead to rather small or disconnected communities. Therefore, network partitioning type of algorithms are more appropriate.

Network partitioning algorithms are further categorized into unified-model methods and separate-model methods based on different strategies they have in handling structure and node information. One solution to network partitioning is through treating the topology structure and node attributes in the same way by a unified model [28], such as a distance metric [25] or a Bayesian probabilistic model [23]. The other solution is through first modelling the topology structure and node attributes separately and then try to combine them to decide the final community structure [24]. The former is referred to as unified model, while the latter is known as separate models.

Separate models are more preferred than unified models. In unified model the algorithms treat structure and node attribute information similarly. This causes loss of information and reduces the capability of the unified-model methods to take full advantage of the partition ability of both types of information. Besides, treating both information in the same way enable unified models to adjust the relative importance of structure and attribute flexibly.

Therefore, an appropriate algorithm to community detection is the one having characteristics of (i) Multi-objectivity (ii) Optimization based and (iii) Separate based model that benefits from both structure and node attribute information. Therefore, in this paper we introduce an evolutionary community detection algorithm called, HarmonY search based Multi-objective Separate model (HYMOS). The algorithm detects useful communities through three stages of harmony search namely as, initialization, improvisation and replacement. The first stage operates once only, and the algorithm iterates between the two other stages of improvisation-replacement.

Evolutionary based algorithms (EAs) divide the solutions into two sets of dominated and non-dominated [17, 21, 22]. According to [17] one of the key challenges of evolutionary based community detection algorithms is how to utilize those non-dominated solutions to enhance the algorithm performance. HYMOS is different from other EA based algorithms by utilizing a subset of non-dominated solutions (called clone solutions) as initial solutions for harmony memory and running the improvisation-replacement cycles, based on that solutions.

The rest of the paper is organised as follows. In Sect. 2 we review some of the most recent EA based algorithms. Section 3 introduces HYMOS in details and finally Sects. 4 and 5 conduct the experiments and draws a conclusion, respectively.

2 Related Works

In this section, we review some of the most recent evolutionary based community detection algorithms to make the differences between HYMOS and the recently proposed algorithms clear. A hybrid multi-objective harmony search based algorithm was proposed by [11]. The proposed algorithm, first applies an improved spectral method to the network, to convert the problem of community detection into a clustering problem. Using an adaptive hybrid multi-objective harmony search algorithm, the multi-objective optimization problem is solved to resolve the community structure.

[22] introduces an algorithm that, combines the structure and attribute information of each node using a Locus based evolutionary algorithm and show that the node information is contributing towards creating more meaningful communities in compare to algorithms that only rely on community structure.

One such algorithm that uses community structure has been proposed by [17]. The evolutionary algorithm has two-phases of community detection and model selection. In the detection phase, MOCD simultaneously optimizes two conflicting and complementary objective functions. Using locus-based adjacency schema, the degree of community partition is measured. A set of community partitions will be supplied to the second phase, where it does trade-off between the two objectives and different numbers of communities. In the model selection phase. Finally, the algorithm selects the most preferable solution from the partition set.

Some authors [7, 8] considered a decomposition based multi-objective evolutionary algorithm proposed by [26] and maximizes the density of internal degrees while minimizes the density of external degrees. In such algorithms, an algorithm with N objective functions will divide the problem into N sub-problems (each correspond to an objective function), and tries to optimize each sub-problem. However, the algorithms of [8] and [7] are different from each other in the way they measure the density of internal degrees.

[1] proposed a multi-objective firefly based community detection algorithm. The algorithm optimizes two objective functions, the community score that measures the density of the clusters obtained and community fitness that minimizes the external links. The proposed algorithm called Enhanced Firefly (EFA). The algorithm Uses a chaotic sequence mechanism to tune the random movement factor with absorption

parameter was set to one. A self-adaptive probabilistic mutation strategy was implemented to improve the convergence characteristic and the quality of the individuals. During the simulation, a set of non-dominated individuals were stored in an external repository, where its size was controlled by a fuzzy-based clustering technique. Using a niching mechanism, the best individuals from the repository are selected. The selection was made in such a way to guide the population towards a smaller search space in the Pareto-optimal front.

3 HYMOS: Proposed Algorithm

In this section, we introduce a harmony search based multi-objective separate model algorithm known as HYMOS for community detection. For more readings on harmony search interested readers may refer to [4]. This algorithm has three main stages of initialization, improvisation and replacement. Each of these steps are explained in details in Sect. 3.1 through Sect. 3.3. In HYMOS we use two objective functions of modularity (Q) and homogeneity (H) that are explained by [22].

3.1 Initialization

Initialization is the first stage of HYMOS, in which dynamic parameters of harmony search are initialized. These parameters are, Number of Iterations (NI) Harmony Memory Consideration Rate (HMCR) and Pitch Adjustment Rate (PAR). Parameter setting of these algorithms are explained in Sect. 4. The other dynamic entity (which is not considered as parameter) is Harmony Memory (HM). Initially HM is filled randomly with values from 1 to total number of nodes in a dataset (k). For the sake of representation we consider a locus-based representation as it meets the requirements of multi-objective community detection [22].

In a locus-based representation an individual (also known as a solution) is represented as a series of integer numbers. In locus-based representation, value v in the ith component vector (i.e., individual) indicates the connection between two nodes of v and i. In other words, two nodes of v and i are in the same community. In Fig. 1, a sample of locus-based representation in HM (see Fig. 1(a)), a harmony vector (Fig. 1 (b)) and decoding of that harmony vector to a community (Fig. 1(c)) are shown. HM has seven columns that corresponds to seven nodes. The initial structure (connection between node and the number of edges) of a community is identified by the dataset.

Then each vector (Fig. 1(b)) will be applied to the initial community and it breaks down the community into a given number of separated communities (Fig. 1(c)). Once the population is initialized, fitness of each vector will be measured using two objective functions of modularity and homogeneity.

3.2 Improvisation

Improvisation is the second stage of HYMOS, in which a new vector will be generated. As we mentioned in the initialization stage, HM is filled randomly. This randomly generated HM does not guarantee proper partitioning (i.e., each HM individual must be able to divide the initial community to at least two sub-communities) of an initial graph. This might also affect the improvisation process, in the sense that no proper community partitioning occurs. We consider this as a shortcoming as it causes generation of useless individuals.

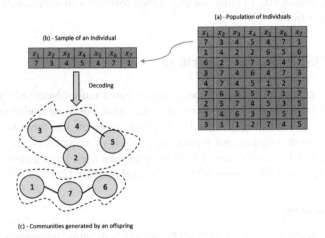

(a) - Population of Individuals

(b) - Sample of an Individual

Decoding

(c) - Communities generated by an offspring

Fig. 1. Locus-based representation of harmony memory and a given harmony vector along with decoding of a harmony vector

Hence, to overcome this shortcoming we assess each individual in HM prior to applying improvisation procedure. This assessment identifies non-dominated (i.e., Pareto front) individuals, and allows using them in improvisation process. Non-dominated individuals are the ones that show superior performances in compare to the other individuals. Therefore, it is more reasonable to improvise new individuals with reliance on non-dominated individuals. In order to identify those non-dominated individuals we apply Pareto Envelope-Based Selection Algorithm II (PESA-II) [2].

PESA-II is a multi-objective optimization algorithm. HYMOS follows the standard principles of an EA with this difference that two populations of solutions are maintained. One is an internal population (IP) of a fixed size, and the other is an external population (EP). The IP explores new solutions through the standard EA process of reproduction and variation. Using EP it is possible to exploit good solutions through maintaining a large and diverse set of the non-dominated solutions discovered during search. The solutions in EP are stored in "niches", in the sense that they are implemented as a hyper-grid in the objective space. A given number of individuals that occupy each niche are kept and used to encourage other individuals to cover the whole objective space, rather than bunch together in one region. Once, PESA-II identified the non-dominated individuals, then HS starts improvisation of a new individual (i.e., harmony vector). A stepwise process of improvisation is shown in Algorithm 1.

Algorithm 1: Improvisation of a new vector

Input:
> k: Number of initial communities in dataset
> P: Total population
> HMCR: harmony memory consideration rate
> PAR: pitch adjustment rate
> *ipsize*: number of internal population
> *epsize*: number of external population

Output:
> NHV: New Harmony Vector

Algorithm:

$P_N \leftarrow$ All Nondominated individuals from P using PESA-II
HM $\leftarrow P_N$

for c $\leftarrow$ 1 to k
    $R_{HMCR} \leftarrow$ generate a random number $U{\sim}(0,1)$
    if $(R_{HMCR} > HMCR)$
        $HM_R \leftarrow$ Randomly choose a row from HM
        $NHV(c) \leftarrow HM_R(c)$
    else
        $NHV(c) \leftarrow U{\sim}(1,k)$
    end
end

3.3 Replacement

Replacement is the third and the last stage of HYMOS. Replacement discards the least significant individual from the population and then inserts the newly generated individual in the population, provided that the new individual is better than the worst individual in the population, in terms of both homogeneity and modularity.

However, since HYMOS is a multi-objective CD algorithm, the question is how to detect the least significant individual in HM and replace it with the newly generated individual. This point should be noted that, replacement is very important, since an improper replacement may cause early or delayed convergence of HYMOS.

As an example, if we only rely on one single objective (either homogeneity or modularity) then the HM will be converged in terms of one single objective (e.g., homogeneity) while the detected communities are suffering in terms of the other objective (e.g., modularity).

In order to help this not happen in HYMOS we consider two different replacement strategies and experimentally investigate their suitability. In one strategy, we first find those individuals in HM that are least significant in terms of homogeneity to the NHV. Then amongst those least significant individuals we replace NHV with the one(s) having modularity lower than NHV's modularity. The other strategy is to swap the first and second objective functions, in the sense that the algorithm first identifies least significant individuals in terms of modularity and then replace NHV with individual(s) having lower homogeneity.

Hence, this introduces two variations of HYMOS. One is known as HYMOS$_{HQ}$ and the other HYMOS$_{QH}$. In HYMOS the loop of improvisation-replacement iterates till the pre-specified number of iterations is met (NI). Algorithm 2 provides a stepwise procedure of HYMOS.

Algorithm 2: HYMOS algorithm

Input:

        k: Number of initial communities in dataset

        P: Total population

        HMCR: harmony memory consideration rate

        PAR: pitch adjustment rate

        ipsize: number of internal population

        epsize: number of external population

        NI: Total number of generations

Output:

        C: Optimal community

Algorithm:

HM $\leftarrow$ initialization(P, k)

g = 1;

while g < NI

$NHV \leftarrow$ Improvisation(k; P; HMCR; HM; PAR; *ipsize*; *epsize*)

    if $(F_{NHV} > F_{HM_{worst}})$ // with respct to replacement strategy

      $HM \leftarrow HM - \{HM_{worst}\} + NHV$

    end

    g = g + 1

end while

$B_g \leftarrow$ Choose the best individual from HM

C $\leftarrow$ decode B_g into a community

3.4 Objective Functions

Communities can be described by two types of information of structure and attribute [21]. The former refers to the connection between nodes (i.e., intra- and inter-connectivity) while the latter refers to similarity of connected nodes in terms of attribute(s). In this paper we consider, structural information and attribute information of communities as two main objective functions, explained in [21].

3.4.1 Structural Information Objective Function

In this paper we utilize modularity as one of the widely used structural objective functions, proposed by Newman [14]. This objective function deals with the dense connection property of communities and can be written Eq. (1):

$$Q(X) = \sum_{g_l \in X} \frac{\sum_{i,j \in g_l} A_{ij}}{2m} - \sum_{g_l \in X} \left(\frac{\sum_{i \in g_l} k_i}{2m} \right)^2 \tag{1}$$

where X is one possible partition, g_l is a cluster of partition X, k_i is the degree of node v_i, and m is the total number of edges in the network.

3.4.2 Attribute Categorization Objective Function

We measure attribute categorization based on Shannon information entropy theory. In this theory, the entropy of a set measures the average Shannon information content of it. A set with high disorder rate have higher Shannon information. This leads to a high entropy, indicating that a given set has high entropy and hence low homogeneity rate. Thus, the entropy-based criterion can be used to measure how homogeneous the elements of a set or a category are. Homogeneity of a community can be measured as Eq. (2):

$$H(X) = \sum_{j=1}^{t} \omega_j H_{b_j}(X) \tag{2}$$

where ω_j the weight of the jth attribute is $H_{b_j}(X)$ is measured through Eq. (3).

$$H_{b_j}(X) = \ln(d_j + 1) - PCE_{b_j}(X) \tag{3}$$

where d_j is the domain of the jth attribute and PCE is measured through Eq. (4).

$$PCE_{b_j}(X) = \sum_{g_l \in X} \frac{n_l}{n} PE_{b_j}(g_l) \tag{4}$$

where g_l is the lth community n_l is the number of edges in the lth community and n is the total number of edges in all the communities. PE_{b_j} is calculated using Eq. (5).

$$PE_{b_j}(X) = -\sum_{q=1}^{d_j} pp_{ij}^q \ln pp_{ij}^q \tag{5}$$

where pp_{ij}^q can be measured using Eq. (6).

$$pp_{ij}^q = \frac{n_{ij}^q}{n_l - 1} \tag{6}$$

Algorithms that produce high rates of homogeneity and modularity are more desired. In Sect. 4 we carry out some experiments to reveal the strengths and weaknesses of HYMOS in compare to its rival algorithms.

4   Experimental Results

In this section, some experiments are carried out to reveal the weaknesses and strengths of HYMOS in compare to two other algorithms of MOCDA [22] and MOCD [17] using two datasets as explained in Sect. 4.1. We empirically fine-tuned HYMOS for two parameters of HMCR and PAR in Sect. 4.2. Number of Iterations (NI) parameter is fine-tuned in Sect. 4.3 by analysing convergence behaviour of HYMOS. Harmony memory Size (HMS) is set dynamically during algorithm execution. Then we studied the convergence behaviour of HYMOS and the competitors, in all the datasets in Sect. 4.3 and finally we compare HYMOS with MOCDA [22] and MOCD [17] in terms of homogeneity and modularity. Further details of these two algorithms are given as below.

- **MOCDA:** This is a multi-objective attributed community detection that relies on evolutionary algorithm of Non-dominated Neighbour Immune Algorithm to detect communities. The procedure divides the population into active and dominant. Further, a subset of population called clone members are extracted from active members to generate new offsprings. The new set of offsprings are then integrated into dominant members to generate the next population for the next iteration.
- **MOCD:** This is a multi-objective evolutionary based community detection algorithm. MOCD optimizes two contradictory objective functions of intra- and inter-connectivity in order to shape communities. MOCD uses PESA-II for the multi-objective component. It maintains two population of IP and EP, where EP exploits good solutions, through maintaining diverse set of the non-dominated solutions found during search.

4.1   Datasets

In the experiments of this section, we use three datasets. We use two datasets of American Football College Network (Football) with 115 nodes and 613 edges [5]. This dataset contains the connections between Division IA colleges. Each node represents a football team and edges between nodes denote the regular games. The second dataset is Books Network (Book) [13] has 105 nodes and 441 edges. This dataset consists of nodes representing books about US politics. Edges represent frequent co-purchasing of books by the same buyers.

4.2   Parameter Tuning

We executed HYMOS with initial population size (i.e., HMS) of 500. HMCR and PAR values are fine-tuned using Books dataset as shown in Figs. 2 and 3, respectively. HMCR experiments in Fig. 2 indicate that each H and Q functions are showing contradictory behaviour to the increase of HMCR from 0.1 to 0.9. H (with some

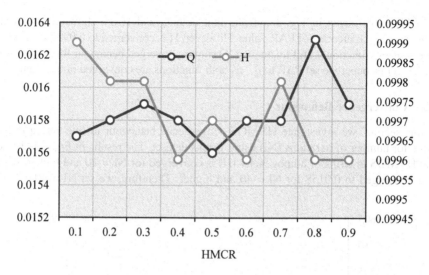

Fig. 2. HMCR fine-tuning using Books dataset

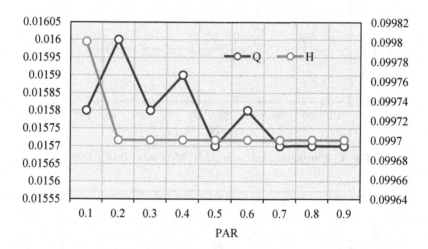

Fig. 3. PAR fine-tuning using Books dataset

fluctuations) has a downward trend while Q (again with fluctuations) has upward trend. Once HMCR = 0.8 then Q is equal to its highest amount of 0.0163 and when HMCR = 0.1 then H is in its peak value.

Therefore HMCR = 0.8 is suitable for Q while HMCR = 0.1 is suitable for H. However, HMCR can have one value at the execution time. Therefore we decided to choose a value for HMCR that both objective functions are showing fair behaviour, (i.e., both functions show upward/downward trend). Since the aim is to maximize each H and Q functions we set HMCR to 0.7, as both H and Q functions are showing upward behaviour in that value.

PAR experiments in Fig. 3 indicate that both Q and H are showing downward behaviour to the increase of PAR value. However, H is converged to 0.0997 for PAR > 0.1, while Q function is still fluctuating. Q shows converged behaviour for PAR > 0.6. Hence in this paper we set PAR to 0.7 as both functions are converged in that value.

4.3 Convergence Behaviour

In this section, we investigate HYMOS convergence behaviour to fine-tune a proper value for number of iterations (NI) using Books dataset. The results in Fig. 4 indicates that HYMOS in terms of homogeneity (H) is converged for NI = 10 and higher, while Q is converged to 0.0158 for NI = 40 and higher. Therefore, we set NI = 40.

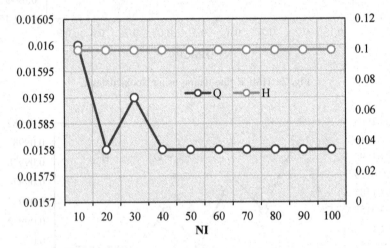

Fig. 4. Convergence behaviour of HYMOS using Books dataset

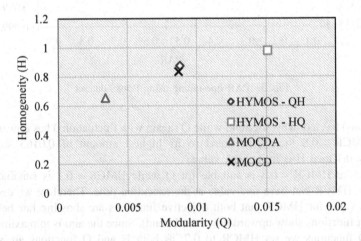

Fig. 5. Homogeneity and Modularity results in Football dataset. Comparisons between HYMOS and two other evolutionary based algorithms

Also PESA-II parameters of ipsize, epsize, crossover probability (pc) and mutation probability (pm) are set separately for each dataset according to the paper of [17]. The Q objective function has converged after 40th iteration while the other objective function is converged after the 10th iteration. This indicates that reliance on HM has significant effect on attribute convergence while an increase in number of iterations can help structure objective function gets converged.

4.4 Performance Comparisons

In this section, we compare HYMOS against MOCDA and MOCD, in terms of homogeneity and modularity. Three datasets of Football, Books and AMT are used and their results are shown in Figs. 5, 6, 7, respectively.

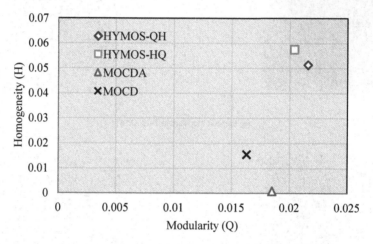

Fig. 6. Homogeneity and Modularity results in Books dataset. Comparisons between HYMOS and two other evolutionary based algorithms

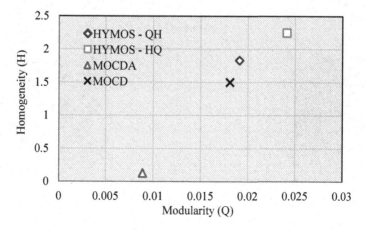

Fig. 7. Homogeneity and Modularity results in AMT dataset. Comparisons between HYMOS and two other evolutionary based algorithms

In both Football and Books datasets HYMOS variations are outperforming
state-of-the-arts in terms of modularity and homogeneity. In Football dataset,
$HYMOS_{HQ}$ is outperforming all other algorithms while in Books dataset $HYMOS_{HQ}$ is
the superior algorithm, merely in terms of Homogeneity. This indicates that HYMOS is
effective enough in outperforming state-of-the-arts, but suitability of replacement
strategy is different from dataset to dataset and it must be investigated.

As we mentioned previously we introduce a new network for AMT workers. The
experiments of this dataset is shown in Fig. 7. Apparently both variations of HYMOS

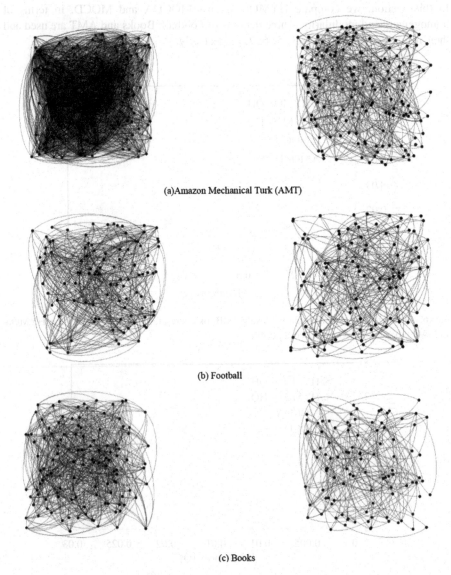

(a)Amazon Mechanical Turk (AMT)

(b) Football

(c) Books

Fig. 8. Detection of communities before (left side) and after (right side) applying HYMOS

are outperforming MOCD and MOCDA. The superiorities gained by HYMOS variations, are due to (i) formulating community detection as a multi-objective meta-heuristic problem using HS and (ii) the utilised replacement strategies, introduced in Sect. 3.3. Figure 8 shows how HYMOS can detect communities of an initial graph. Apparently, the detected communities are overlapping.

5 Conclusion and Future Works

In this paper, we introduced HYMOS as one of the multi-objective evolutionary based algorithm for community detection. HYMOS relies on HS to detect communities. HYMOS initially generates a random population and then using PESA-II algorithm estimates those, non-dominated individuals. Using those non-dominated individuals the algorithm generates a given number of NHVs. each generated NHV, will be replaced with a member of HM using two strategies.

One is to first select members that are outperformed by NHV in terms of modularity and then replace NHV with those outperformed in terms of homogeneity. The other is to first select those members outperformed in terms of homogeneity and then replace NHV with those members that are also outperformed in terms of modularity.

We experimentally showed the utility and efficacy of HYMOS by using three datasets. The experiments showed that HYMOS is able to outperform recent state-of-the-art algorithms in terms of homogeneity and modularity. This is resulted from meta-heuristic search power of HS and the utilised strategies in replacing a NHV. As future works one can overcome the shortcoming of evolutionary based CD algorithm which is entrapment in local optima, by hybridizing HYMOS with a proper local search.

References

1. Amiri, B., et al.: Community detection in complex networks: multi–objective enhanced firefly algorithm. Knowl.-Based Syst. **46**, 1–11 (2013)
2. Corne, D.W., et al.: PESA-II: Region-based selection in evolutionary multiobjective optimization. In: Proceedings of the 3rd Annual Conference on Genetic and Evolutionary Computation (2001)
3. Fortunato, S.: Community detection in graphs. Phys. Rep. **486**(3), 75–174 (2010)
4. Geem, Z.W.: Novel derivative of harmony search algorithm for discrete design variables. Appl. Math. Comput. **199**(1), 223–230 (2008)
5. Girvan, M., Newman, M.E.: Community structure in social and biological networks. Proc. Natl. Acad. Sci. **99**(12), 7821–7826 (2002)
6. Gong, M., et al.: Identification of multi-resolution network structures with multi-objective immune algorithm. Appl. Soft Comput. **13**(4), 1705–1717 (2013)
7. Gong, M., et al.: Community detection in networks by using multiobjective evolutionary algorithm with decomposition. Phys. A: Stat. Mech. Appl. **391**(15), 4050–4060 (2012)
8. Hariz, W.A., Abdulhalim, M.F.: Improving the performance of evolutionary multi-objective co-clustering models for community detection in complex social networks. Swarm Evol. Comput. **26**, 137–156 (2016)

9. Li, S., et al.: Detecting community structure via synchronous label propagation. Neurocomputing **151**, 1063–1075 (2015)
10. Li, T., Ma, S., Ogihara, M.: Entropy-based criterion in categorical clustering. In: Proceedings of the Twenty-First International Conference on Machine Learning (2004)
11. Li, Y., et al.: A spectral clustering-based adaptive hybrid multi-objective harmony search algorithm for community detection. In: IEEE Congress on Evolutionary Computation (CEC) (2012)
12. Moser, F., et al.: Mining cohesive patterns from graphs with feature vectors. In: Proceedings of the SIAM International Conference on Data Mining (2009)
13. Newman, M.E.: Modularity and community structure in networks. Proc. Natl. Acad. Sci. **103** (23), 8577–8582 (2006)
14. Newman, M.E., Girvan, M.: Finding and evaluating community structure in networks. Phys. Rev. E **69**(2), 026113 (2004)
15. Pool, S., Bonchi, F., van Leeuwen, M.: Description-driven community detection. ACM Trans. Intell. Syst. Technol. TIST **5**(2), 28 (2014)
16. Sese, J., Seki, M., Fukuzaki, M.: Mining networks with shared items. In: Proceedings of the 19th ACM International Conference on Information and Knowledge Management (2010)
17. Shi, C., et al.: Multi-objective community detection in complex networks. Appl. Soft Comput. **12**(2), 850–859 (2012)
18. Shi, C., et al.: On selection of objective functions in multi-objective community detection. In: Proceedings of the 20th ACM International Conference on Information and Knowledge Management (2011)
19. Shi, C., et al.: Comparison and selection of objective functions in multiobjective community detection. Comput. Intell. **30**(3), 562–582 (2014)
20. Vitali, S., Battiston, S.: The community structure of the global corporate network. PLoS ONE **9**(8), e104655 (2014)
21. Wu, P., Pan, L.: Multi-objective community detection based on memetic algorithm. PLoS ONE **10**(5), e0126845 (2015)
22. Wu, P., Pan, L.: Multi-objective community detection method by integrating users' behavior attributes. Neurocomputing **210**, 13–25 (2016)
23. Xu, Z., et al.: A model-based approach to attributed graph clustering. In: Proceedings of the ACM SIGMOD International Conference on Management of Data (2012)
24. Yang, J., McAuley, J., Leskovec, J.: Community detection in networks with node attributes. In: IEEE 13th International Conference on Data Mining (ICDM) (2013)
25. Zhang, H., et al.: Semi-supervised distance metric learning based on local linear regression for data clustering. Neurocomputing **93**, 100–105 (2012)
26. Zhang, Q., Li, H.: MOEA/D: A multiobjective evolutionary algorithm based on decomposition. IEEE Trans. Evol. Comput. **11**(6), 712–731 (2007)
27. Zhou, Y., Cheng, H., Yu, J.X.: Graph clustering based on structural/attribute similarities. Proc. VLDB Endowment **2**(1), 718–729 (2009)
28. Zhou, Y., Cheng, H., Yu, J.X.: Clustering large attributed graphs: an efficient incremental approach. In: 2010 IEEE 10th International Conference on Data Mining (ICDM) (2010)

A Semi-supervised Hidden Markov Topic Model Based on Prior Knowledge

Sattar Seifollahi[1,2(✉)], Massimo Piccardi[1], and Ehsan Zare Borzeshi[2]

[1] Faculty of Engineering and Information Technology,
University of Technology Sydney, Sydney, NSW, Australia
sseif@cmcrc.com
[2] Capital Markets Cooperative Research Centre (CMCRC),
Sydney, NSW, Australia

Abstract. A topic model is an unsupervised model to automatically discover the topics discussed in a collection of documents. Most of the existing topic models only use bag-of-words representations or single-word distributions and do not consider relations between words in the model. As a consequence, these models may generate topics which are not in good agreement with human-judged topic coherence. To mitigate this issue, we present a topic model which employs topically-related knowledge from prior topics and words' co-occurrence/relations in the collection. To incorporate the prior knowledge, we leverage a two-staged semi-supervised Markov topic model. In the first stage, we estimate a transition matrix and a low-dimensional vocabulary for the final topic model. In the second stage, we produce the final topic model where the topic assignment is performed following a Markov chain process. Experiments on real text documents from a major compensation agency demonstrate improvements of both the PMI score measure and the topic coherence.

Keywords: Topic modelling · Hidden Markov model
Latent Dirichlet allocation · Topic coherence

1 Introduction

Topic models are a frequently used tool in text mining. They are commonly referred to as probabilistic topic models to refer to all probabilistic algorithms that reveal the hidden structure of a document collection. In essence, a topic model can be seen as an algorithm that takes a collection of documents as its input and returns its set of topics. In addition to explaining the collection, topic models have been widely employed as a low-dimensional representation of the documents in information retrieval, document summarization, and classification [2,4,13].

The motivation for topic models goes back to the latent semantic indexing (LSI) [6]) which is a concept model originally introduced for information

© Springer Nature Singapore Pte Ltd. 2018
Y. L. Boo et al. (Eds.): AusDM 2017, CCIS 845, pp. 265–276, 2018.
https://doi.org/10.1007/978-981-13-0292-3_17

retrieval. The earliest topic model as such was introduced by Papadimitriou et al. [15] and a significant step in topic modelling progress was due to [9,10] who introduced a probabilistic model for the LSI (pLSI). Latent Dirichlet allocation (LDA) [2] is an extension of pLSA and is the most widely used topic model. LDA considers the standard bag-of-words (BoW) representation, where each document is explained as a mixture of topics, with each topic describing a probability distribution over the words in the collection.

Most of the current topic models are extensions of LDA such as the Pachinko allocation, [19] which introduced topics' correlations further to the word correlations to form the topics. Examples of other extensions include, but are not limited to, hierarchical formulations to produce an unknown number of topics [17], topic evolution over time [3,18,21], topic correlation using the logistic normal distribution [1], topics that follow a Markov chain [16] and change over sentences [8], and topics employing prior knowledge [5,11,12,20].

Although the LDA model has been the basis of many models, one of its limitations and of many related models is that they do not consider subsequent words' relations in the model. In [8], topics of sentences follow a Markov chain. They assume that topics' changes may occur between sentences, and that the change in topic is decided at the beginning of the sentence by drawing from a binomial distribution. However, we believe that such an assumption may be restrictive and may lead to incoherent topics in challenging corpora.

The aim of this research is to develop a topic model based on both the LDA model and how words topically relate. To do this effectively, we present a two-staged semi-supervised hidden Markov topic model in which topics are assigned to individual words, regardless of their sentences, based on how each word is topically connected to the previous word in the collection. The first stage uses a standard LDA model to find a large, initial number of "prior" topics. These prior topics not only give us words' relations in the collection and how they are topically related, but also provide us with a low-dimensional representation of the vocabulary to use in the second stage of the model. Therefore, the model uses a semi-supervised learning technique leveraging such prior topics and the co-occurrences of words in the collection. The present work is different from all existing models, and especially from the original LDA model and the hidden topic Markov model introduced in [8], in that the topic assignment follows a semi-supervised Markov chain rather than a random walk. This makes our model intrinsically more capable of capturing word relations within topics. In general, a Markov chain is a fundamental assumption for encapsulating the relationship between successive events, and in topic modelling it can be used over topics, documents, sentences, and, herewith, words.

The rest of our paper is structured as follows: Sect. 2 provides a brief discussion on LDA. In Sect. 3 we present our methodology in detail. Evaluation of algorithms and discussions are reported in Sect. 4. Finally, the conclusions are drawn.

2   Latent Dirichlet Allocation

The LDA model, [2], is the most widely used probabilistic topic model to derive a low-dimensional topic space. It assumes the BoW representation of the words, or exchangeability assumption, which is common to most topic modelling approaches. Each document d is represented as a discrete distribution over topics, where each topic k, at its turn, is a discrete distribution over the words. Both distributions, the document-topic θ_d and topic-words ϕ_k, are drawn from Dirichlet distributions with parameters α and η, respectively. Figure 1(a) illustrates the graphical model of LDA, where D is the number of documents, N the number of words, and K the assumed number of topics in the collection. The documents are known, whereas the topic parameters - including the document-topic, topic-words and topic assignments - are unknown or hidden. The generative model equivalent to the graphical model of LDA 1(a) is presented in Algorithm 1. With the above notations and generative process, the joint probability of a document d with N words can be formulated as:

Algorithm 1. Latent Dirichlet Allocation

---

1. For each topic k in $\{1, \ldots, K\}$.
   1.1. Draw a distribution over words, $\phi_k \sim Dir_W(\eta)$.
2. For each document d :
   2.1. Draw a vector of topic proportions, $\theta_d \sim Dir_K(\alpha)$.
   2.2. For each word $w_{d,n}$
      2.2.1. Draw a topic assignment, $z_{d,n} \sim Mult_K(\theta_d)$.
      2.2.2. Draw a word, $w_{d,n} \sim Mult_W(\phi_{z_{d,n}})$.

$$p(w) = \int_\theta \Big(\prod_{n=1}^{N} \sum_{z_n=1}^{K} p(w_n|z_n; \phi) p(z_n|\theta) \Big) p(\theta; \alpha) d\theta \tag{1}$$

The posterior computation based on (1) is highly intractable; therefore the need for an estimation method is crucial. There are generally two types of estimation methods; (i) the variational Bayesian estimation and (ii) the (collapsed) Gibbs sampling. The expectation maximization (EM) algorithm is usually used to learn the LDA parameters by maximizing a variational bound; see [17], while the aim in Gibbs sampling, [7], is to estimate the posterior distribution by a Markov chain that converges to the limiting distribution. The Markov chain runs for a large number of iterations, collects samples from the conditional distributions, and then approximates the distribution with the obtained samples. Herewith, we make use of the collapsed Gibbs sampling.

3   Semi-supervised Markov Topic Model

In this section, we discuss our topic modelling approach in detail. The graphical representation of the model is displayed in Fig. 1(b). As shown in this figure, it

consists of two stages; (i) prior topics and (ii) final topic model. The aim of the
first stage is to determine a subset of prior topics to calculate words' relations
upon. More precisely, a transition matrix for the topic assignment in the second
stage is calculated by looking at both subsequent words in the prior topics and
co-occurrences of words in the collection. The frequent co-occurring words in the
collection and in prior topics represent the more topically-related words - this is
the main idea behind the proposed topic model, called semi-supervised hidden
Markov topic model, or SHMTM, hereafter. In addition, the prior topics will
also be used to extract a low-dimensional vocabulary for the final model.

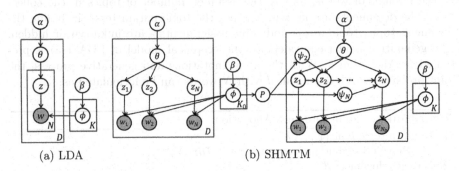

(a) LDA   (b) SHMTM

Fig. 1. Graphical models of (a) LDA model and (b) the proposed model, SHMTM.
The number of prior topics in SHMTM is K_0 ($K_0 > K$) and the vocabulary size is N_0
($N_0 \leq N$)

3.1 Initial Process

Frequently co-occurring words in collections can be much more topically-related
than words without significant co-occurrence. To calculate words relations, we
first extract a topic model with a high number, K_0, of topics using the LDA
model. For each pair of words in the vocabulary, or in a lower-dimensional vocabulary using only the top words, we calculate the probability of words co-occurring
in the collection given the prior topics. The words relations give us a measure
to find how two subsequent words are topically related. This measure, called a
"transition matrix" herewith for analogy with Markov chains, is used by the final
topic model for topic assignment. The initialisation procedure can be summed
up as follows: The number of prior topics, K_0, is equal to or greater than the
final number of topics, and there is no specific criterion for an optimal number
of prior topics. K_0 can be considered as an upper bound for the possible topics
and it may depend on the data set and application; we fix $K_0 = 200$. The number of top words n_k is also fixed to 20 based on the fact that only a few words
from each topic have significant impact on the topics. The term $P_{i,j}$ in Step 4,
is proportional to the probabilities of the i-th and j-th words in the collection
where w_i and w_j are both top words in the same topic; they are calculated as:

Algorithm 2. Initial topic model

---

1. Set the number of prior topics, K_0.
2. Find K_0 topics using the LDA model.
3. Select the top n_k words of each topic with high probabilities and set $W_0 = \{w_i^k\}; i = 1, \ldots, n_k; k = 1, \ldots, K_0$.
4. Calculate a transition probability matrix, $P = [P_{i,j}]_{i,j=1}^{N_0}$, between top words of each topic, where N_0 is the size of set W_0.

$$P_{i,j} = \sum_k p(w_j^k, w_i^k) p(w_j^k), \tag{2}$$

where $p(w_j^k, w_i^k)$ is the probability of co-occurrence of words w_j and w_i in the collection (or related external data), and $p(w_j^k)$ is the probability (weight) of word w_j in the k-th topic. We calculate the score for any two words in the vocabulary by investigating the top n_k words of each topic. If two words do not appear as top words of any topics, a weight of zero will be assigned to the corresponding score, meaning that those words are not topically related (or independent given the topic). In contrast, if two words appear in many topics and frequently co-occur in the collection, they are more likely to be dependent words or topically-related words.

To leverage (2) as a transition probability matrix, one need to scale each row of P to interval $[0, 1]$. The diagonal of this matrix is also set to 1 to ensure that two (subsequent) equal words are assigned to the same topic. As P is not symmetric, i.e. $P_{i,j} \neq P_{j,i}$, one can consider the mean of P and the transpose of P, which is used in this paper. It is noted that only top words of prior topics are used to construct a transition matrix as well as a low-dimensional vocabulary for the final model.

3.2 Generative Model

We propose our topic model where the topic assignment is formed in a semi-supervised way, i.e. based on a transition matrix obtained in the previous step. The central idea is that highly topically-related, subsequent words are generated from the same topics. However, the extent to which subsequent words are topically related is decided based on a Markov chain process. A transition matrix is first constructed using both prior topics and word co-occurrences in the collection, and topic assignment is then decided using the transition matrix and a random number drawn from a binomial distribution. The steps of the generative model are given in Algorithm 3. As shown in Step 2.3.1. of Algorithm 3, the topic of each word is decided based on two factors: (i) the transition matrix and (ii) the binomial distribution. The topic of the first word in each document follows the standard LDA topic assignment, i.e., it is generated from a Dirichlet distribution. The model can be regarded as semi-supervised overall since the process is a Markov chain employing prior knowledge.

Algorithm 3. Semi-supervised hidden Markov topic model (SHMTM)

1. For each topic k in $\{1, \ldots, K\}$.
  1.1. Draw a distribution over words, $\phi_k \sim Dir_{W_0}(\eta)$.
2. For each document d :
  2.1. Draw a vector of topic proportions, $\theta_d \sim Dir_K(\alpha)$.
  2.2. For $n = 1$:
    2.2.1. Draw a topic assignment, $z_{d,1} \sim Mult_K(\theta_d)$.
    2.2.2. Draw a word, $w_{d,1} \sim Mult_{W_0}(\phi_{z_{d,1}})$.
  2.3. For $n = 2, \ldots, N_d$:
    2.3.1. Draw a binomial random number, $b_n \sim binomial(P_{n-1,n})$.
    2.3.2. If $b_n = 0$, set $z_{d,n} = z_{d,n-1}$, else draw a new topic $z_{d,n} \sim Mult_K(\theta_d)$.
    2.3.3. Draw a word, $w_{d,n} \sim Mult_{W_0}(\phi_{z_{d,n}})$.

4 Experiments

4.1 Data Set

We present experiments based on two data sets from the Transport Accident Commission (TAC) which is a major accident compensation agency of the Victorian Government in Australia. The first data set is a collection of 593,433 phone calls from 13,937 single TAC clients recorded by various operators over 5 years. The phone calls are made for different purposes including, but not limited to, compensation payments, recovery and return to work; different type of services, medications and treatments; pain, solicitor engagement and mental health issues. The second data set is a collection of file notes about 6,991 TAC clients, taken by TAC managers over the same period to create new "tasks" for the clients according to their health outcomes, income, etc. We refer to the first data set as "phonecalls", for short, and the second data set as "filenotes". Both data sets require a large amount of preprocessing including: (1) removal of numbers, punctuation, symbols and "stopwords" from the collection; (2) synonyms and misspelled words are replaced with the base and actual words; (3) infrequently occurring words are removed from the collection; (4) as common in text mining, we also remove the most frequently occurring words based on a predefined list such as names and addresses. Finally, the collection of documents is converted into sequences of numeric features.

4.2 Results and Evaluation

We use the point-wise mutual information (PMI score) to evaluate the proposed model using: (i) a low-dimensional vocabulary of top words from prior topics and (ii) the whole vocabulary of the collection. According to [14], the PMI score is a suitable measure for comparing topics' quality and coherence. For ease of reference, hereafter we refer to the SHMTM using the top words of prior topics in the dictionary as SHMTM-top, the SHMTM using all words of the document collection for the dictionary as SHMTM-all, the LDA with top words of prior topics as LDA-top, and the LDA with all words of the collection as LDA-all. All algorithms were implemented in Python 3.5. The PMI score is proportional to the probability of co-occurrence of words normalised by the probabilities of individual words [14], and is expressed as:

$$PMI(k) = \frac{1}{45} \sum_{i<j} PMI(w_i, w_j), \ i, j \in \{1, \ldots, 10\}, \tag{3}$$

where

$$PMI(w_i, w_j) = \log \frac{P(w_i, w_j)}{P(w_i)P(w_j)}$$

where $P(w_i, w_j)$ is the probability of co-occurring i-th and j-th words in the collection, and 45 is the number of distinct possible word pairs in the top 10 words. The above Eq. (3), measures the PMI score of the k-th topic, and the PMI score of K topics is simply the average of their PMI score values. Figure 2 shows how words are topically related for some words using 200 prior topics on the phonecalls data set. The higher the value, the more topically related. Based on this table, "return to work" (rtw) is highly related to other words including "recovery" and "progress"; hence, their topic assignments will very likely be the same.

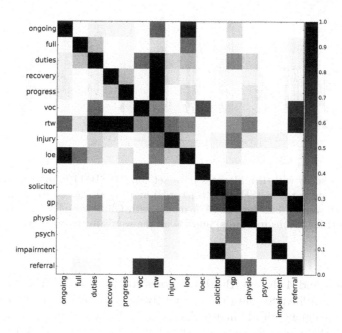

Fig. 2. A sample set of words and their relations in the transition matrix. The relations have been created using 200 prior topics on the phonecalls data set.

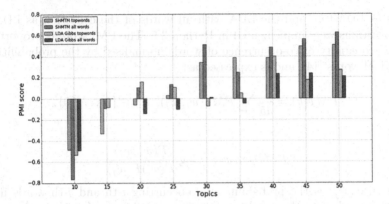

Fig. 3. PMI score with the phonecalls data set for the SHMTM model using (1) the top words of prior topics (SHMTM topwords) and (2) all the words in the collection (SHMTM all words), and for the LDA model with Gibbs sampling using (3) the top words only (LDA Gibbs topwords) and (4) all the words in the collection (LDA Gibbs all words)

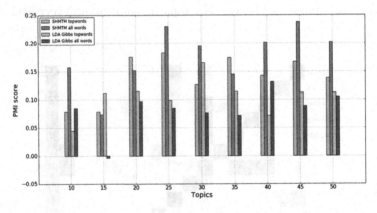

Fig. 4. PMI score with the filenotes data set for the SHMTM model using (1) the top words of prior topics (SHMTM topwords) and (2) all the words in the collection (SHMTM all words), and for the LDA model with Gibbs sampling using (3) the top words only (LDA Gibbs topwords) and (4) all the words in the collection (LDA Gibbs all words)

Figures 3 and 4 display the PMI scores for LDA and SHMTM using the phonecalls and filenotes data sets, respectively. The higher the PMI score, the more the agreement with human-judged topic coherence [14]. Based on the PMI scores with the phonecalls data set, SHMTM performs better than LDA, particularly when the number of topics is greater than 20. In addition, the PMI scores with the filenotes data set show improvement in most cases. These figures also illustrate the variation of the PMI score with the number of topics. With the phonecalls data set, the PMI scores get more stable when the number of topics gets to around 30; for the filenotes data set, around 20.

Tables 1 and 2 present the top words for four topics (out of a total of 35) of each model with the phonecalls data set. Although some topics of the different models look similar in terms of top words, words of topics in SHMTM-all and SHMTM-top are more topically-related than from the other two models. For example, words "training" and "retraining" appear as top words in the first topic of LDA-all, but the topic also contains unrelated words such as "solicitor", "loec" (loss of earning compensation) and "impairment" as other top words. This is also shown by Fig. 3, where the SHMTM-all model proves to be the most accurate among all models (for numbers of topics greater than 20), followed by SHMTM-top, LDA-top and LDA-all.

Table 1. Top words from four topics for SHMTM using only the top words or the whole dictionary with the phonecalls data set

SHMTM-top			
gp	work	voc	taxi
loec	rtw	employment	travel
report	back	job	work
solicitor	pain	appointment	drive
psych	recovery	voc_provider	approval
attend	treatment	work	appointment
treatment	physio	job_seeking	physio
impairment	gp	pay	driving
appointment	weeks	income	transport
decision	long	attend	assist
SHMTM-all			
solicitor	rtw	voc	taxi
loec	treatment	job	travel
impairment	recovery	work	work
report	holistic	employment	ankle
provider	followup	voc_provider	rtw
gp	back	job_seeking	drive
assessment	time	loec	foot
decision	physio	service	unable
psych	gp	voc_service	approval
cease	work	provider	physio

Table 2. Top words from four topics for LDA using only the top words or the whole dictionary with the phonecalls data set

LDA-top			
gp	rtw	voc	taxi
report	treatment	loec	travel
solicitor	assist	job	medcert
impairment	recovery	voc_provider	work
injury	time	employment	appointment
back	support	capacity	review
med	left	work	drive
decision	holistic	job_seeking	hospital
psych	gp	service	rtw
time	issue	voc_service	transport

LDA-all			
solicitor	pain	loec	taxi
pay	treatment	voc	Travel
impairment	rtw	employment	service
back	psych	income	home
report	recovery	job	approval
psych	week	voc_provider	appointment
cso	weeks	work	assist
followup	gp	capacity	drive
retraining	back	job_seeking	support
jme	holistic	cease	weeks

5   Conclusion and Future Work

In this paper, a semi-supervised hidden Markov topic model has been proposed using prior knowledge. The model consists of two stages where the prior knowledge is accounted for in the first stage and coherent topic models are produced by the final model. In the model's assumptions, the topic of a word not only follows a Dirichlet distribution, but also depends on the topic of the previous word. The topic dependency is decided using a binomial distribution and a transition probability matrix obtained by an initial modelling of prior topics. Experiments on real text documents (clients phone calls and file notes) from the Transport Accident Commission of the Victorian Government in Australia have given evidence to improvements of the PMI score measure and topic coherence.

Acknowledgement. This project was funded by the Capital Market Cooperative Research Centre in combination with the Transport Accident Commission of Victoria. Acknowledgements and thanks to industry supervisors David Attwood (Lead Research

Partnerships) and Bernie Kruger (Data Science Lead). This research has received ethics approval from University of Technology Sydney (UTS HREC REF NO. ETH16-0968).

References

1. Blei, D.M., Lafferty, J.D.: A correlated topic model of science. Ann. Appl. Stat. **1**(1), 17–35 (2007)
2. Blei, D.M., Ng, A.Y., Jordan, M.I.: Latent Dirichlet allocation. J. Mach. Learn. Res. **3**, 993–1022 (2003)
3. Blei, D.M., Lafferty, J.D.: Dynamic topic models. In: International Conference on Machine Learning (ICML) (2006)
4. Blei, D.M., McAuliffe, J.D.: Supervised topic models. In: Advances in neural information processing systems, vol. 20, pp. 121–128. MIT Press, Cambridge (2008)
5. Chen, Z., Liu, B.: Topic modeling using topics from many domains, lifelong learning and big data. In: Proceedings of the 31st International Conference on Machine Learning (ICML), vol. 32, pp. 703–711 (2014)
6. Deerwester, S., Dumais, S.T., Furnas, G.W., Landauer, T.K., Harshman, R.: Indexing by latent semantic analysis. J. Assoc. Inf. Sci. Technol. **41**(6), 391–407 (1990)
7. Griffiths, T.L., Steyvers, M.: Finding scientific topics. Proc. Nat. Acad. Sci. **101**, 5228–5235 (2004)
8. Gruber, A., Rosen-Zvi, M., Weiss, Y.: Hidden topic Markov models. In: Proceedings of the 11th International Conference on Artificial Intelligence and Statistics (AISTATS-07), pp. 163–170 (2007)
9. Hofmann, T.: Probabilistic latent semantic indexing. In: Proceedings of the 22nd Annual International ACM SIGIR Conference on Research and Development in Information Retrieval, pp. 50–57 (1999)
10. Hofmann, T.: Unsupervised learning by probabilistic latent semantic analysis. Mach. Learn. **42**, 177–196 (2001)
11. Hsu, W.S., Poupart, P.: Online Bayesian moment matching for topic modeling with unknown number of topics. In: Proceedings of the 30th International Conference on Neural Information Processing Systems (2016)
12. Hu, Y., Boyd-Graber, J., Satinoff, B.: Interactive topic modeling. Proceedings of the 49th Annual Meeting of the Association for Computational Linguistics, pp. 248–257 (2011)
13. Lacoste-Julien, S., Sha, F., Jordan, M.I.: DiscLDA: discriminative learning for dimensionality reduction and classification. In: Proceedings of the 21st International Conference on Neural Information Processing Systems, pp. 897–904 (2008)
14. Newman, D., Bonilla, E.V., Buntine, W.L.: Improving topic coherence with regularized topic models. In: Proceedings of the 24th International Conference on Neural Information Processing Systems (2011)
15. Papadimitriou, C.H., Raghavan, P., Tamaki, H., Vempala, S.: Latent semantic indexing: a probabilistic analysis. J. Comput. Syst. Sci. **61**(2), 217–235 (2000)
16. Qiang, J., Chen, P., Wang, T., Wu, X.: Topic modeling over short texts by incorporating word embeddings. In: Kim, J., Shim, K., Cao, L., Lee, J.-G., Lin, X., Moon, Y.-S. (eds.) PAKDD 2017. LNCS (LNAI), vol. 10235, pp. 363–374. Springer, Cham (2017). https://doi.org/10.1007/978-3-319-57529-2_29
17. Teh, Y.W., Newman, D., Welling, M.: A collapsed variational Bayesian inference algorithm for latent Dirichlet allocation. In: Proceedings of the 19th International Conference on Neural Information Processing Systems, pp. 1353–1360 (2006)

18. Wang, C., Blei, D.M., Heckerman, D.: Continuous time dynamic topic models. In: Proceedings of the 24th Conference on Uncertainty in Artificial Intelligence (2008)
19. Wei, L., Blei, D.M., McCallum, A.: Nonparametric Bayes pachinko allocation. In: Proceedings of the 23rd Conference on Uncertainty in Artificial Intelligence, pp. 243–250 (2007)
20. Wood, J., Tan, P., Wang, W., Arnold, C.: Source-LDA: enhancing probabilistic topic models using prior knowledge sources. In: IEEE 33rd International Conference on Data Engineering (ICDE) (2017)
21. Xuerui, X., McCallum, A.: Topics over time: a non-Markov continuous-time model of topical trends. In: Proceedings of the 12th ACM SIGKDD International Conference on Knowledge Discovery and Data Mining, pp. 424–433 (2006)

Author Index

Author Index